T0234724

SpringerBriefs in Physics

Series Editors

Balasubramanian Ananthanarayan, Centre for High Energy Physics, Indian Institute of Science, Bangalore, India

Egor Babaev, Physics Department, University of Massachusetts Amherst, Amherst, MA, USA

Malcolm Bremer, H H Wills Physics Laboratory, University of Bristol, Bristol, UK

Xavier Calmet, Department of Physics and Astronomy, University of Sussex, Brighton, UK

Francesca Di Lodovico, Department of Physics, Queen Mary University of London, London, UK

Pablo D. Esquinazi, Institute for Experimental Physics II, University of Leipzig, Leipzig, Germany

Maarten Hoogerland, University of Auckland, Auckland, New Zealand

Eric Le Ru, School of Chemical and Physical Sciences, Victoria University of Wellington, Kelburn, Wellington, New Zealand

Dario Narducci, University of Milano-Bicocca, Milan, Italy

James Overduin, Towson University, Towson, MD, USA

Vesselin Petkov, Montreal, QC, Canada

Stefan Theisen, Max-Planck-Institut für Gravitationsphysik, Golm, Germany

Charles H.-T. Wang, Department of Physics, The University of Aberdeen, Aberdeen, UK

James D. Wells, Physics Department, University of Michigan, Ann Arbor, MI, USA

Andrew Whitaker, Department of Physics and Astronomy, Queen's University Belfast, Belfast, UK

SpringerBriefs in Physics are a series of slim high-quality publications encompassing the entire spectrum of physics. Manuscripts for SpringerBriefs in Physics will be evaluated by Springer and by members of the Editorial Board. Proposals and other communication should be sent to your Publishing Editors at Springer.

Featuring compact volumes of 50 to 125 pages (approximately 20,000–45,000 words), Briefs are shorter than a conventional book but longer than a journal article. Thus, Briefs serve as timely, concise tools for students, researchers, and professionals.

Typical texts for publication might include:

- A snapshot review of the current state of a hot or emerging field
- A concise introduction to core concepts that students must understand in order to make independent contributions
- An extended research report giving more details and discussion than is possible in a conventional journal article
- A manual describing underlying principles and best practices for an experimental technique
- An essay exploring new ideas within physics, related philosophical issues, or broader topics such as science and society.

Briefs allow authors to present their ideas and readers to absorb them with minimal time investment.

Briefs will be published as part of Springer's eBook collection, with millions of users worldwide. In addition, they will be available, just like other books, for individual print and electronic purchase.

Briefs are characterized by fast, global electronic dissemination, straightforward publishing agreements, easy-to-use manuscript preparation and formatting guidelines, and expedited production schedules. We aim for publication 8–12 weeks after acceptance.

More information about this series at http://www.springer.com/series/8902

James D. Wells

Discovery Beyond the Standard Model of Elementary Particle Physics

 Springer

James D. Wells
Physics Department
University of Michigan
Ann Arbor, MI, USA

ISSN 2191-5423   ISSN 2191-5431   (electronic)
SpringerBriefs in Physics
ISBN 978-3-030-38203-2   ISBN 978-3-030-38204-9   (eBook)
https://doi.org/10.1007/978-3-030-38204-9

© The Author(s), under exclusive license to Springer Nature Switzerland AG 2020
This work is subject to copyright. All rights are reserved by the Publisher, whether the whole or part of the material is concerned, specifically the rights of translation, reprinting, reuse of illustrations, recitation, broadcasting, reproduction on microfilms or in any other physical way, and transmission or information storage and retrieval, electronic adaptation, computer software, or by similar or dissimilar methodology now known or hereafter developed.
The use of general descriptive names, registered names, trademarks, service marks, etc. in this publication does not imply, even in the absence of a specific statement, that such names are exempt from the relevant protective laws and regulations and therefore free for general use.
The publisher, the authors and the editors are safe to assume that the advice and information in this book are believed to be true and accurate at the date of publication. Neither the publisher nor the authors or the editors give a warranty, expressed or implied, with respect to the material contained herein or for any errors or omissions that may have been made. The publisher remains neutral with regard to jurisdictional claims in published maps and institutional affiliations.

This Springer imprint is published by the registered company Springer Nature Switzerland AG
The registered company address is: Gewerbestrasse 11, 6330 Cham, Switzerland

Preface

The Standard Model (SM) of elementary particle physics has been a central theory in high-energy physics since the early 1970s. However, there are many reasons to believe that it is incomplete, including its lack of a good dark matter candidate, its inability to explain the matter-antimatter asymmetry, and its incompleteness to solve the problems inflation sets out to solve. Furthermore, the SM has little to say about the intriguing prospects for grand unification that the gauge coupling values and their converging renormalization group trajectories suggest. Nor does it have any understanding for the hierarchy of masses within the theory, most notably the dramatically smaller masses for neutrinos than for other particles in the spectrum, and the lightness of the weak scale in the face of (possible) destabilizing Planck-scale dynamics, motivated by many scenarios that solve the problems stated above.

Yet, to make discoveries beyond the SM (BSM) and answer these questions is not a straightforward task. In addition to the difficulty of securing resources (materials and people) necessary to do additional experiments and additional theory work, there is a question of exactly how one should go about searching for new physics. Should we just do much more of the same (intensity frontier) to get more and more precise measurements to compare with the SM? Or, should we turn up the energies on our colliders, likely requiring building more colliders, to search into the energy frontier? Should we chase theories that theorists say are the best and brightest theories, or should the experiment be agnostic about theories? These are the questions that I pursue in Chap. 1, where I advocate that a mutual theory-experiment effort where both work together in the search for new physics is vastly superior, and even logically more sound, than working separately. In other words, a BSM-oriented approach, born of this cooperation between experiment and theory, is argued to be more efficacious in our search for new physics than signal-based only approach ("signalism").

In Chap. 2, I remind the reader that BSM physics is a murky topic without understanding what the "Standard Model" really means. I argue that the "Standard Model" should not be considered a static moniker, and that our conception of it has always been a little hazy, including today, and that conception has changed repeatedly since the birth in the early 1970s of the "Standard Model" as the preeminent recognized theory of elementary particles. I also argue that it is important to clear up the hazy definition, which requires us to recognize that the SM is not just a collection of facts and rules, but also has inherited mysteries and myths when it faces the task of explaining all the natural phenomena within its ostensible purview.

The main hope of this volume is not necessarily to convince the reader of my particular viewpoints on these questions, but rather to familiarize the reader with the issues so that they will develop their own considered opinion. The choices and direction of the field, especially at this juncture in time, depend on its researchers reflecting deeply on what are the pathways that will best lead to discovery, continuing our productive search for a deeper understanding of the laws of nature.

Ann Arbor, USA   James D. Wells
December 2019

Acknowledgements I wish to thank many collaborators and colleagues who have been influential to me in thinking on these issues, some broadly and some very specifically for this work. They include R. Akhoury, D. Amidei, S. de Bianchi, G. Ballesteros, G. Giudice, G. Kane, S. P. Martin, and A. Pierce. I would like to thank the Humboldt Foundation, the Leinweber Center for Theoretical Physics (LCTP), and the U.S. Department of Energy for support during the time this volume was written.

Contents

Chapter 1
Discovery Goals and Opportunities: A Defense of BSM-Oriented Exploration over Signalism

Abstract Discoveries come through exclusions, confirmations or revolutionary findings with respect to a theory canon populated by the Standard Model (SM) and beyond the SM (BSM) theories. Guaranteed discoveries are accomplished only through pursuit of BSM exclusion/confirmation, and thus require investment in the continual formation and analysis of a vibrant theory canon combined with investment in experiment with demonstrated capacity to make BSM exclusions or confirmations. Risks develop when steering away from BSM-oriented work toward its methodological rival, "signalism," which seeks to realize SM falsification or revolutionary discoveries outside the context of any BSM rationale. It is argued that such an approach leads to inscrutable exertions that reduce prospects for all discovery. The concepts are applied to the European Strategy Update, which seeks to identify future investments in forefront experiment that bring a balance of guaranteed and prospective value.

1.1 Introduction

The practice of science includes a wide range of activities, ranging from theoretical speculations to experimental analysis. These activities are all in the pursuit of scientific discovery—securely knowing something of science value that we did not know before. In this essay a formalization of the language of discovery is put forward that articulates common ambient notions in high-energy physics. From this, an argument is made that persistent and guaranteed discovery, as well as enhanced prospects for discovery of every kind, are accomplished through the co-work of beyond the Standard Model (BSM) theory and experimental work focused on BSM exclusion/confirmation. Signalism is the main methodological rival to BSM-centered exploration. It proposes to achieve SM falsification or revolutionary discoveries without any reference to BSM theories. However, it will be argued that signalism is an inscrutable and non-rational methodology for science discovery and puts at risk all types of discovery as conventionally conceived.

The thesis introduced above should not be interpreted to imply lower value or lesser status for other activities such as SM theory work, SM experimental analysis,

© The Author(s), under exclusive license to Springer Nature Switzerland AG 2020
J. D. Wells, *Discovery Beyond the Standard Model of Elementary Particle Physics*,
SpringerBriefs in Physics, https://doi.org/10.1007/978-3-030-38204-9_1

formal theory, or detector/experimentation development. These, as we shall see, are indispensable activities ultimately in the service of discovery when done well. Nevertheless, it is argued that positioning BSM as the central attractor of theory work and experiment discovery is what guarantees, vindicates, and gives meaning to those other efforts.

This essay is admittedly long. The impatient reader can go straight to the summary (Sect. 1.10) to read a listing of the main points developed. The full essay aims to give context, justification and nuance to those claims. Sections 1.2–1.5 set up the conceptualization of discovery, with arguments and illustrations for the BSM-centered approach peppered throughout. Section 1.6 addresses the methodological rival "signalism" more directly, and suggests that it comes up short compared to BSM-centered work. In some sense, Sect. 1.6 is the culmination of the main thesis of the essay that BSM-centered work is superior to signalistic approaches for the pursuit of discovery. Sections 1.7 and 1.8 illustrate the main points of the essay through discussion of recent discoveries of gravity waves and the Higgs boson, and also through discussion of the European strategy update, which aims to make possible more discoveries in the future. Section 1.9 discusses the risks and signs of discovery ending, and their antidotes. Section 1.10 summarizes the essay.

1.1.1 Theoretical Versus Experimental Discovery

Let us continue the introduction by first discussing a little more on what is meant by "discovery" in this essay. Colloquially we refer to discoveries mainly within the experimental realm. There are exceptions, such as speaking of Einstein having discovered General Relativity, whereas Eddington only confirmed it experimentally, or rather discovered a unique predicted feature of the theory (bending of light). However, the majority of cases where the appellation discovery is applied is reserved to experimental work: Thomson discovered the electron; Rutherford discovered the proton; Chadwick discovered the neutron; Anderson and Neddermeyer discovered the muon; Richter's and Ting's collaborations discovered the J/ψ; the Gargamelle collaboration discovered neutral currents; the CDF and D0 collaborations of Fermilab discovered the top quark; Atlas and CMS collaborations of CERN discovered the Higgs boson; etc.

Standard usage of discovery in science rightly puts the primary emphasis on experiment. Applying the word "discovery" for the invention of a theory, whenever it does happen, as in the case of General Relativity, often only takes place after experimental confirmation, which is the strongest form of discovery. Discovery has the sense of uncovering something that is true that was lying in wait for us to find. For us, theories will not be evaluated in our forthcoming discussion on whether they always existed or whether they are permanently true fixtures waiting to be found, but rather whether they are presently adequate in the face of all experimental results known. Thus, it would be preferable perhaps to replace phrases such as "she discovered the theory of X" with something less provocative, such as "she educed the theory of X".

Colloquially we may also utilize the word "discovery" for the product of a "founder of discursivity," as a Foucauldian might say, where a new work produces "the possibilities and the rules for the formation of other texts" [63], where "other texts" in our context are forthcoming scientific works made possible by the founding work. A key example of this in recent years was the "discovery" of warped extra dimensions by Randall and Sundrum [92], which resulting in a multitude of additional works that built upon their founding idea. When Randall and Sundrum were appropriately awarded the 2019 Sakurai Prize[1] their citation was for "in particular the discovery that warped extra dimensions of space can solve the hierarchy puzzle..." [96]. The word "discovery" is implicitly modified by "theoretical" by the context of the award being exclusively in the theoretical domain. However, there has been no experimental verification (not yet at least) of warped extra dimensions. Therefore, by the common implicit rules of scientific discourse one could not say in a contextless environment that "warped extra dimensions have been discovered." Only after experimental verification could one presume to make such a grand statement. For this reason, the unmodified word "discovery" in a contextless sentence must necessarily refer to a result confirmed by experiment, such as "the discovery of general relativity", or "the discovery of neutrinos," etc.

Nevertheless, theory plays a significant role in the discovery process. Many times experimental discoveries are made because they constructed dedicated apparatuses to search in subtle places that theory suggested. The most celebrated recent example of that is the discovery of the Higgs boson, which required a multi-billion dollar experiment with special particle detectors designed primarily with the Higgs boson discovery requirements in mind. Thus, any full accounting of discovery must also make theory an integral part of the story.

1.1.2 Experiment as Transformations of the Theory Canon

The key construct through which we account for theory's role in discovery is what can be called the *theory canon*. The theory canon is the collection of all theories devised, including the standard reference theory (i.e., the Standard Model in particle physics), that satisfy all the requirements that physicists believe make these theories good descriptions of nature. There are many such requirements. Some are uncontroversial (i.e., must satisfy all known experimental data, must be mathematically consistent), while others are controversial (i.e., must be natural, must be simple, must not be in swampland). It is not just theorists who decide what belongs in the canon, but all stakeholders that test such theories. For this reason what is admitted into the theory canon is a difficult community discussion.

---

[1] The Sakurai Prize is the highest award given by the American Physical Society for work in theoretical particle physics.

More will be said about the theory canon later, but let us suppose we have one. Experimental discoveries are then made within the context of that canon. Confirmations are made when a theory or a key component of a theory within the canon is confirmed. Exclusions are made with respect to a theory in the canon. (One cannot exclude what one does not know.) Similarly, relegation or falsification of a theory to the dustbin of history (i.e., total exclusion) is an experimental discovery that can only be achieved if there is a theory canon within which the falsified theory had once lived. The existence of the theory canon enriches experiment and makes possible numerous discoveries that were otherwise inconceivable.

Of course, there are experimental discoveries that take place completely outside the context of the theory canon. Finding completely unexpected particles or interactions or signals that are unanticipated by any theory within the theory canon is revolutionary. Such revolutionary discoveries (e.g., discovery of the muon is thought to be one such discovery) are part of physics history and presumably should continue to be into the future. Nevertheless, it should be noted that what makes them spectacular, eye-popping, revolutionary and rare is the existence of an advanced theory canon that is exploded by the discovery.

In the following, three broad categories of experimental discovery are described: confirmation, exclusion, and revolutionary. There are important further distinctions and subcategories that will be made within these broad categories, which includes SM confirmations, BSM confirmations, falsifications of the SM, falsification of a BSM theory, or falsification of the entire theory canon. As stated at the top, it will be argued that perpetual and guaranteed discovery passes through the focusing gateway of BSM theory and BSM-centered experiment.

1.1.3 The Work of Assured Discovery

It is hoped that articulating the concepts, categories and paths of discovery will contribute to assessing valuable activity in high-energy physics enterprise, especially as we plan for its future. As we contemplate all the aspects of guaranteed discovery, we see that the effort that gives rise to it can be organized into three core discovery activities that must be healthy for high-energy physics to be healthy:

- "model building": constructing a vibrant and motivated BSM theory canon.
- "theory analyzing": connecting theory canon ideas with phenomenological implications.
- "experimental work": translating phenomenological implications of the theory canon into experiments with assured confirmation/exclusion capability.

All three of these are necessary, and require intense, focused and unique knowledge and skill sets.

The categories above are based on action-oriented work, not static labels of individuals, since a scientist can in principle participate in any combination of these three activities, although he/she most often has hard-won primary expertise in only one. Often a physicist can have substantial overlap in nearest-neighbor activities. For example, a physicist can contribute to "model building" and "theory analyzing" and yet another can contribute to "theory analyzing" and "experimental work". It is hoped that it will become clear after the argument is presented that if any of these activities dwindles, guaranteed discovery ends.

The above list may give the reader the wrong impression that more formal theoretical work is viewed here as less relevant to discovery and less important. Formal work includes many areas of active research including string theory, AdS/CFT theory, amplitudes theory, finite temperature field theory, black hole conundrumology, information theory, etc. Although formal work looks far removed from discovery it is recognized by most to contribute as feed-in fuel to model building and theory analysis. For example, the proof of renormalizability of the weak interactions was critical to progress in concretizing the SM into a fully calculable theory (see Veltman's and 't Hooft's essays in [73]). As another example, the work of dualities in supersymmetric Yang-Mills theory [99] led to significant developments in BSM model building [47]. Likewise, AdS/CFT correspondence [80] has given much deeper and fruitful correspondences between theories of warped extra dimensions and walking technicolor [15], which on the surface looked unconnected. The recent development in the theory of amplitudes (for reviews, see, e.g., [43, 59]) is hoped to one day provide a significantly better approach to theory analysis, and perhaps even model building. Similarly, in the past, the mathematical physics work of group theory, topology, differential geometry, etc. also could not have been spoken of directly as "model building" or "theory analyzing" as discussed above, yet they ultimately have played central roles in both.

One could then interpret formal work as vital "pre-discovery" work in the service of model building and theory analysis which is, in turn, in the service of (experimental) discovery. It is no less an important activity as any of the others for a healthy and vibrant field that wishes to continue making discoveries far into the future. Nevertheless, if a particular activity of formal physics cannot be plausibly argued to have some possible connection to the three more direct discovery activities (model building, theory analyzing, experimental work), then it is at risk of being a less relevant activity. It is a subtle task to evaluate formal theory work's ultimate relevance to discovery. That topic will be taken up elsewhere. For the purposes of this essay we need merely acknowledge that the "pre-discovery" work of formal physics is crucial and contributes fuel to sustained progress in model building and theory analysis.

Lastly, just as formal work within theory gives fuel to future advances in model building and theory analyzing, so does "pre-discovery" work in experimental physics. Detector R&D, accelerator physics research, computational and electronics hardware advances, and analysis software tools, all contribute toward and seed progress in experimental work. In some sense this is the experimental analogue to theory's "formal work," which is less direct and proximate to actual discoveries, but is vital work that enables more direct discovery activities to realize themselves consistently into the future.

1.2   The Theory Canon

One of the aims of theoretical physics is to seek theories that have predictive capacity and are empirically adequate. Empirical adequacy is the ability of at least one point of the theory's parameter space to match all experimental measurements simultaneously within a stipulated domain of applicability. If there exists one point in parameter space that is empirically adequate there usually exist an infinite number of point that are empirically adequate—a "good" region of parameter space. For example, in the SM there are an infinite number of input parameter points that match the data, such as the infinite number of top quark mass values in the experimentally allowed range 172.26 ± 0.61 GeV [101].

There can be numerous theories that are consistent with all known data. For example, in addition to the SM there is the minimal supersymmetric standard model [81], the next-to-minimal supersymmetric standard model, the minimal composite Higgs theory [85], the $SU(2)$ left-right gauge theory model, the minimal warped extra dimension model [92], the large extra dimension model [14], the SM Effective Theory (SMEFT) theory with higher dimensional operators [34, 35], etc. All of these are called beyond the Standard Model (BSM) theories. Each of these theories has its own parameter space and may have different extended domains of applicability[2] beyond the minimal domain required for the SM success.

The collection of all theories that are empirically adequate are candidates for admission into the theory canon. Certainly the SM is within the theory canon, since it is the agreed-upon standard reference theory that agrees with the data. In addition the SM, the theory canon contains the union of every empirically adequate BSM theory that a non-trivial subset of the expert scholarly community deems to have value beyond the SM. In addition to the SMEFT, the manifesting practice is to admit new theories into the canon that "explain more" than the SM, such as dark matter, fermion mass and mixing angle hierarchies, small Higgs mass, small cosmological constant, coexistence with gravity, origins of spacetime symmetries and structure, origins of internal symmetries, coexistence with unification, baryon asymmetry of the universe, etc.

It must be repeatedly emphasized that every theory within the theory canon must be at present consistent with all known experiment. Furthermore, every theory within the canon is there provisionally and is never safe. Additional theory analysis can find that a theory is incompatible with an experiment in a way that had not before been understood. Or, an experiment can release a new experimental result that ejects theories from the canon. The parameter spaces of theories within the canon are continually under revision. The theory canon shape-shifts often based on theoretical and experimental progress. Occasionally it is even annihilated when new revolutionary

---

[2]By extended domains of applicability it is meant that a theory may purport to have a definite range of validity, such as a minimal supersymmetric theory up to the grand unification scale. Or, it may have augmented purposes compared to the SM, such as providing a dark matter candidate. This is the case of new theory that looks like the SM except it has, for example, one more real scalar S that couples to the Higgs boson and is postulated to be the dark matter of the universe [51].

experimental results are inconsistent with every theory within the canon. In that case, a new theory canon is reborn, perhaps slowly, in the wake of this revolutionary development.

A key concept put forward in this work is that experimental discovery can and should be understood and categorized in terms of the effect it has on the theory canon. The above discussion of the theory canon and its relevance to experiment is somewhat abstract and general by constructive necessity, but the implications are very concrete and the recognitions of various types of discovery are straightforward. In the following sections various forms of discovery are defined with respect to the theory canon, with specific examples provided to give greater clarity and applicability to the abstract notions. But first, we must say a word more about the standard reference theory (i.e., the SM of particle physics) and how it is represented in the full theory canon space.

As mentioned above, the categories of discovery that are discussed in more detail are delineated by the action that experiment does on the theory canon. It is useful to have a visualization of these various actions. In order to do that we must develop a visual representation of the theory canon based on the principles discussed earlier of what is in the canon.

We can visualize the theory canon as a collection of all admitted BSM theories, each of which is represented by a parameter space where the allowed region is demarcated (for us, in green), as seen in Fig. 1.1. Here, the BSM theory has parameters η_1 and η_2 such that when $\eta_i \to 0$ all observables reproduce SM values. Thus, $\eta_{1,2}$ are decoupling parameters of a decoupling BSM theory, which is currently the most representative type of BSM theory within the theory canon. The BSM theory may have many more parameters than just these two graphically shown, but the concept is the same: The SM is represented as a limiting point at the origin. Non-decoupling theories may not have a SM point anywhere in the visualization, but for it to be in the theory canon it must have experimentally allowed regions of parameter space. An important feature of a non-decoupling theory is that it can be ruled out even if the SM is exact in nature.

To take advantage of this decoupling behavior in visualizing a specific BSM theories, it is helpful to recast the BSM parameters such that the SM decoupling limit is always at the origin. In other words, instead of plotting $m_{1/2}$ versus m_0, where the SM limit is really the "point at infinity", we construct the inverse as parameters, e.g., $\eta_{1/2} = m_Z/m_{1/2}$ and $\eta_0 = m_Z/m_0$, where the origin of the $(\eta_{1/2}, \eta_0)$ is the decoupling limit of the SM.[3] One can then represent the allowed region more compactly, and scientific progress and discovery is a tighter push the origin with confirmation discovery potential ever present.

---

[3]Minimal supersymmetry is not exactly a decoupling theory in the sense that the Higgs mass is computable in terms of superpartner masses and is not a free to be any value in the low-scale SM effective theory. For this reason, the allowed parameter space will never include exactly the origin in the $(\xi_0, \xi_{1/2})$ parameter space, or equivalently at the point at infinity in the $(m_0, m_{1/2})$ parameter space, as illustrated for example by the allowed region of Fig. 1.1 of [23] being restricted to finite values in the $(m_0, m_{1/2})$ plane.

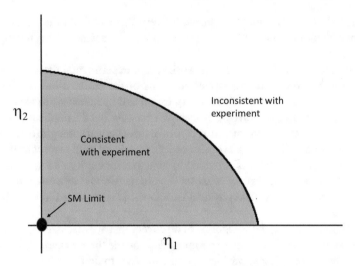

Fig. 1.1 Green is the currently allowed parameter space of a BSM theory that decouples to predictions for observables indistinguishable from those of the SM when $\eta_{1,2} \to 0$

Another example is "sequential hypercharge Z'" theory. This is a new Z' boson that couples to the SM in exactly the same way as the hypercharge gauge boson except that its mass $M_{Z'}$ and overall coupling strength $g_{Z'}$ are free parameters. This is arguably not within the theory canon since its motivation may not be high enough, but it demonstrates visualization in an especially simple way, which is analogous to many theories that are within the theory canon, such as dark photon dark matter theories. The ξ-variables for this representation are $g_{Z'}$ and $m_Z/M_{Z'}$ which gives decoupling (i.e., SM predictions) at the origin. It is also true that the entire $g_{Z'} = 0$ and $m_Z/M_{Z'} = 0$ axes are in the decoupling limit as well. It is this reason that experimental constraints on the parameter space will push closer and closer to the $g_{Z'} = 0$ and $m_Z/m_{Z'} = 0$ axes but can never get there. As the exclusion capability increases it nevertheless does open the opportunity for a signal to develop in that previously unexplored region of parameter space. That would constitute an important discovery.

As alluded to above, there may be other theories that have no decoupling limit at all to the SM. For example, the case of minimal no-scale supergravity theories with neutralino dark matter LSP do not allow superpartner masses to decouple [55]. This theory is similar to the standard minimal constrained supersymmetric standard model except that $m_0 = 0$ is required, which puts an upper bound on $m_{1/2}$, otherwise the LSP is no longer a neutralino and so cannot be the dark matter. The upper bound on $m_{1/2}$, and thus lower bound in $\xi_{1/2} = m_Z/m_{1/2}$, prevents reaching a decoupling limit within the theory.

One might object that minimal no-scale supergravity is just a subset of the parameter space within the more expansive minimal supergravity theories that do not require $m_0 = 0$ and thus should not be consider as an additional theory within the theory canon. However, landmarks of experimental progress are powerfully stated as total exclusion of coherent, self-contained BSM theories with specifically motivated theoretical structures and phenomenological targets (such as dark matter, $g - 2$ explanation, etc.). Recognitions of BSM falsifications are powerful milestones, which can in turn also impact views of the community on the larger category of theories, such as whether minimal supersymmetry should still occupy high table in the theory canon.

We have discussed minimal supersymmetry and minimal Z' models within this discussion of the theory canon. But there are many more ideas of high interest to the high-energy physics community including warped extra dimensions, twin Higgs theories, little Higgs theories, minimal scalar-extended dark matter scenarios, superlight vector dark matter, low-scale baryogenesis sector theories, etc. The stature of various theories within the theory canon is not the subject of this essay, yet it must be recognized that various ideas are promoted and others relegated as their strengths and weakness are revealed in the intense theoretical and experimental scrutiny they experience. In this sense there is value in some "group-think" activity to promote, criticize and explore ideas. A thousands scientists in a thousand attics working on a thousand totally distinct ideas are unlikely to make the progress needed for discovery. Likewise, a thousand scientists in one attic working on only one idea is also unlikely to engender a healthy flow of ideas and discovery. As with most such endeavors, a balance between these two extremes toward constructing and analyzing the theory canon is likely to most useful. However, balance in this sense is not to be recommended to exist within every individual, but rather across the field, since individuals must focus to make impact. Partly for this reason, banishing an idea from the theory canon that has many invested proponents is not easy. Nevertheless, theory ideas die regularly, albeit it quietly with few visiting the graves (minimal technicolor, minimal non-supersymmetric $SU(5)$ GUTs, supersymmetric electroweak light-stop baryogenesis, minimal conformal SM, etc.).

Let us also remark that it is entirely reasonable to be cautious of theory talk about such lofty aims as "elucidating the true structure of space and time" and "constructing deeper reformulations of the laws of nature," etc. A new, improved language is not particularly transformative if one cannot order a good dinner with it, as every speaker of Esperanto can attest. Less controversial is a more instrumentalist appraisal of knowledge gain and theory development, which assesses the ability to predict that "if I do A, then I know B will come next", where, of course, B can be a collection of probabilistic outcomes. This power of prediction is worth more than any fancy subtle theory or "deep insight" into the soul of nature.[4] However—and this can never be forgotten—powerful workhorse predictive theories are often given birth by lofty theory/mathematical parents (e.g., non-abelian gauge symmetries, general coordinate invariance, supersymmetric theories, conformal theories, etc.). Thus, erring on the side of inclusive acceptance to theory development is in order, but researchers in theoretical high-energy physics should be able to articulate how their work is (or

---

[4]Distinguishing true science from mere visionary pronouncements has been a difficult problem for millinia. Nevertheless, as scholars frequently note, "we have come to realize that the best proof that our knowledge is genuine is that it enables us to do something" [61].

at least "might be") connected to the construction of new BSM theories that answer outstanding problems in nature (i.e., ability to make predictions or to explain "histories"), or they should be able to explain how their work enables (or at least "might enable") more effective analysis of the SM and BSM canon theories that enlarges capacity for exclusion/confirmation discoveries. Theory work that can do neither is unlikely to contribute to genuine discovery.

Finally, our purpose here is not to develop an evaluative theory of what should and should not be in the theory canon, or a praxis theory of how some theories get promoted and others banished among empirically adequate alternatives, or other such philosophical concerns. The purpose here is mainly to point out that a theory canon does indeed exist, as any high-energy physicist recognizes. They are the theories that many people continue to work on. They are the theories that experimentalists aim to find or constrain. They are the theories that end up in technical design reports motivating new experiments. Furthermore, the theory canon exists even though individual physicists might differ on what the community views as being contained within it, especially some theories on the "edges" of the canon (somewhat fewer practitioners, less experimental interest, remaining allowed parameter space is extremely "small" compared to prior motivated assessments, etc.). Criticisms, promotions, additions and deletions of the theory canon will always be a part of high-energy physics. Nevertheless, discovery is and should be made with respect to that canon, as will be developed more fully below. These discoveries are confirmation, exclusion and revolutionary, to which we now turn.

1.3 Confirmation Discoveries

With respect to the theory canon, there are three kinds of confirmation discoveries. The first kind of confirmation, SM feature confirmation, is experimentally verifying a feature of the SM that hitherto had not yet been established, or even was viewed by many as highly uncertain. The second type of confirmation, SM locus confirmation, is confirming by experiment the empirical adequacy of a narrowly carved locus of points in SM parameter space motivated by additional principles that go beyond the SM definition (i.e., BSM motivated). And a third type of confirmation discovery, BSM confirmation, is verifying a feature of a BSM theory by which the SM is eliminated from the theory canon and the BSM theory is elevated to the new SM.

1.3.1 SM Feature Confirmation

Let us first consider a SM confirmation. Throughout the history of particle physics there are many such examples. Notable ones in recent years include discoveries of the charm quark [18, 20], of the W boson in 1983 [16, 25], the top quark in 1995 [7, 8], and the Higgs boson in 2012 [3, 41]. The charm quark and Higgs boson discoveries were particularly momentous since confidence that they should be found was not uniform among high-energy physicists. In addition to finding evidence for these elementary particles, a SM confirmation discovery can be said to involve any qualitative property or manifestation of the theory that had not yet been observed. Examples of these include the presence of CP violation in B decays [9, 19], discovered in 2001, observation of CP violation in charm decays [5], discovered very recently, and the existence of three active species of light neutrinos [29].

The determination of three neutrino species was partially achieved before the start of LEP's Z-pole experiments in 1989, where a review then noted that data was consistent with $N_\nu = 2.0^{+0.6}_{-0.4}$, and concluded that "$N_\nu = 3$ is perfectly compatible with all data. Although the consistency is significantly worse, four families still provide a reasonable fit. In the framework of the Standard Model, a fifth light neutrino is, however, unlikely" [54]. Very quickly after the turn-on of LEP and SLC, measurements of invisible final states of the Z width suggested 3.12 ± 0.19 as of October 1989 [29]. By the time a final analyses were being completed on the precision electroweak data at LEP/SLC the precision completely ruled out anything but 3 neutrino species "assuming that only invisible Z decays are to neutrinos coupled according to SM expectations" [98]. In other words, the SM feature of three neutrinos had been confirmed.

Every SM confirmation discovery is momentous since it both signifies a leap in experimental sophistication and it secures knowledge that we could not be sure of before. In addition, it expels speculations (i.e., BSM theories) that certain features of the SM are indeed absent or altered in the true underlying theory. For example, disbelief and dramatic alternatives to the SM Higgs explanation of electroweak symmetry breaking and mass generation thrived within the theory canon up to the moment of the Higgs discovery [106], underscoring the importance of SM confirmation discoveries.

One of the subtleties about a SM confirmation discovery resides in the meaning of confirmation. Confirmation colloquially often implies that one thing (theory, fact, etc.) was found to be true beyond any doubt and all relevant alternatives are not ("I have confirmed that Thurston attended the opera last Wednesday night.") This is too restrictive of a notion for confirmation in scientific research that aims to go beyond "for all practical purposes" for those satisfied with the needs of the here and now and who feel no compelling desire to delve deeper. However, our task is to continually refine our explanations for physical phenomena. Confirmation for us then is achieving a strong localization within the theory canon, without requiring that the localization returns one and only one empirically adequate theory.

This subtlety regarding confirmation reared its head in the early days of the Higgs discovery when CERN scientists were hesitating calling what they found a "Higgs boson." Instead, they came up with other phrases such as "new particle consistent with the Higgs boson" [3, 40] and "new particle ... with spin different than one" [41]. The reason for this is that there were initially many other theories in the canon (dilaton scalars, spin-2 resonances, etc.) that could have explained the measurements they were getting at those early stages. As the CMS collaboration concluded, "The results presented here are consistent, within uncertainties, with expectations for a standard model Higgs boson. The collection of further data will enable a more rigorous test of this conclusion" [41]. As the data accrued and some of the more exotic ideas (e.g., heavy graviton-like objects) were becoming more inconsistent with the measurements, CERN scientists felt more and more comfortable in 2014, nearly two years after its first discovery, to simply declare that "it has been identified as a Higgs boson" [42].

Nevertheless, precisely what does it mean to say the Higgs boson has been discovered? If it means a scalar boson that has all the decay branching partial widths of the textbook SM Higgs boson field to six significant digits, then nobody can say we have confirmed that. Experimental uncertainty combined with the existence of many ideas within the theory canon that can give small deviations within experimental allowances forbid us from declaring with certainty that what we label as the Higgs boson in the SM is indeed what has been discovered.[5] Instead, more precisely we have discovered a narrowed localization in the theory canon consistent with the existence of a scalar boson and consistent with all the properties of the SM Higgs boson to within measurement errors. Some properties have not been measured to within even an order of magnitude of its predictions (e.g., triple Higgs coupling) whereas other properties have been measured to within about 15% (e.g., Higgs decays to photons and Z bosons) [17, 46].

Despite the caveats, one does now hear said that "the Higgs boson has been confirmed," with the implication that this SM feature has been confirmed. This simple proclamation is acceptable since as a community we know that it is short-hand for "events have been registered in the detector that are consistent with what would be created by the existence of a SM Higgs boson, and whose precision measurements are sufficient to highly suggest that indeed something rather close the SM Higgs boson was found if not the SM Higgs boson itself, although measurements in even the near term might force us to relabel the object again as a Higgs-like particle that shares properties with the SM Higgs but does not quite have exactly SM Higgs properties due to small deviations measured in its couplings to other states compared to those derived from precision SM analysis." Such implicit Joycean statements are the bane of all confirmation discoveries, but they do reveal more accurately the nature of confirmations, whose exasperating tentativeness rewards us nevertheless with the seeds of future possible discovery.

1.3.2 SM Locus Confirmation

The parameters of the standard theory are never measured perfectly. For example, a parameter may be known to within a factor of two, and later measured to within 1% after dedicated experimental study. Whenever there are uncertainties in the parameters there remains the prospect of hypothesizing a higher structure on the standard

---

[5]Indeed, several future colliders, such as ILC [22, 82], HL-LHC and HE-LHC [39] and CLIC [10] are being proposed to discover BSM theories that give altered Higgs boson phenomena in subtle ways [50].

theory that predicts what those parameters converge to when a much better measurement is made later. Or there might be a relation between parameters that is required by assuming an additional symmetry structure on top of the minimal symmetries required to define the standard theory. A "SM locus confirmation discovery" is when, upon further experimental improvements, a BSM theory-derived locus of points in the SM parameter space is measured to indeed be the experimentally selected region.

One way that a locus prediction arises is when analyzing a motivated BSM theory that is in full flourish only at higher energy scales where the new particles and forces have their characteristic scales. Upon integrating out this BSM theory and proceeding to a lower energy SM effective theory, the constraints of the full theory may lead to a tight restriction on what values the SM parameters may take. A graphical illustration of this is given in Fig. 1.2 where an experimentally allowed region in the parameter space of two SM variables c_1 and c_2 is shown in green. The SM treats c_1 and c_2 as independent variables, but the BSM theory predicts that the relation between c_1 and c_2 is fixed and given by the black curved line in the figure. If future experiment makes significant progress and new measurements localize at the locus points of interest, this is a locus confirmation discovery. Figure 1.3 schematically depicts this type of discovery.

The challenge with a locus confirmation discovery such as that depicted schematically in Fig. 1.3 is that the confirmation might not survive additional theory or experiment scrutiny. On the theory side, it is always the case that calculations might not have been complete or correct, and the locus of points identified were incorrectly positioned. Such errors are simply errors and must be rectified, just as there is the possibility of experiment making error. More subtle is when to apply the label "con-

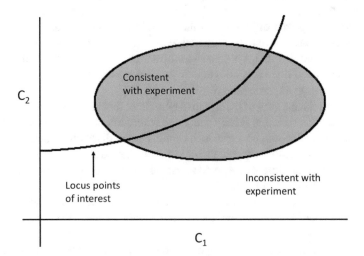

Fig. 1.2   The green region shows the allowed region by all current experiment of two SM parameters C_1 and C_2. The solid black line is a "locus of interest" from the point of view of BSM theory that reduces to the SM in the low-energy limit. This BSM prediction can then be tested by future experiment

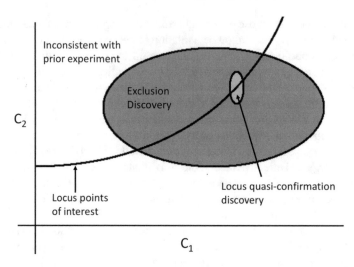

Fig. 1.3 After further experiment, much of the (C_1, C_2) parameter space of the SM is excluded except for a small remaining green region that is centered on the "locus points of interest" predicted by a BSM theory. This is only a quasi-confirmation discovery since there are points allowed by experiment that nature could select that are off the locus points of interest line

firmation" even if every thing were done correctly by theorists and experimentalists. In Fig. 1.3 the black line of locus points is thinner than the extent of the new green experimentally allowed region. Although the experimental focussing on this a priori established locus of points is very impressive, it is also possible that future experiment will result in a significantly smaller green allowed region that does not overlap with the locus of points of interest on the black line. In such a case, the notion of "confirmation" would have to be retracted. For this reason, one may wish to call the discovery depicted in Fig. 1.3 a "quasi-confirmation discovery" rather than a confirmation discovery, since it is not guaranteed by any means that the discovery will hold up after further experimental results.

On the other hand, if the transformative experiment results in a new green region of experimentally points that is fully contained within a continuous locus of points of interest, that indeed would be a true locus confirmation discovery. Such a discovery would not be subject to new categorization from new experiments in the future, unless of course experiments had made mistakes of a systematic nature. This kind of discovery is depicted in Fig. 1.4.

Locus confirmation discovery does not mean that the BSM theory that gave rise to that locus has been "confirmed" or "discovered." Only the locus has been confirmed. This can create increasing interest in the BSM theory and lead to investigations on how other more direct tests, involving qualitatively new phenomena, may be devised that could lead to a more direct BSM confirmation discovery, as will be discussed in Sect. 1.3.3.

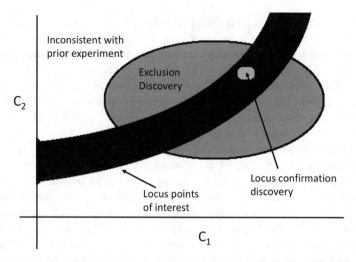

Fig. 1.4 After further experiment, much of the (C_1, C_2) parameter space of the SM is excluded except for a small remaining green region that is centered on the "locus points of interest" predicted by a BSM theory. This is a locus confirmation discovery since the experimentally allowed region is entirely within the locus points of interest

There are numerous instances of locus confirmation discoveries in recent high energy physics. Minimal supersymmetry with supersymmetry scale less than a TeV predicted that the Higgs boson mass should be less than about 135 GeV before the Higgs boson was found.[6] In this case, the locus of points of interest was all masses below 135 GeV for the lightest scalar Higgs bosons. The discovery of the existence of the Higgs boson, as predicted and in opposition to other Higgsless theories, and second that its mass was 125 GeV, thus less than 135 GeV, is considered by many an important success of the theory. Nevertheless, although it is an interesting locus confirmation discovery, it is obviously not to be considered a supersymmetry confirmation discovery, as we emphasized in more general terms above.

1.3.3 BSM Confirmation

The third kind of confirmation discovery, BSM confirmation, occurs when experiment excludes the SM (the reference theory) and localizes around the parameter space of a BSM theory. Many BSM theories could be consistent with the new-found localization within the theory canon, and so many BSM theories could rightly lay

---

[6]See Martin's discussion [81] on p. 54 of version 1 from 1997 which put the upper limit on MSSM light CP-even Higgs mass at $\lesssim 130$ GeV and then on p. 95 of version 4 from 2011 (just prior to Higgs boson discovery), which put the upper limit at $\lesssim 135$ GeV from improved supersymmetric Higgs mass calculations.

claim to a confirmation discovery. What is key is that the SM is excluded and at least one BSM theory in the theory canon remains empirically adequate.

To schematically represent a BSM confirmation discovery, we revert back to our BSM parameterization illustration of two variables η_1 and η_2 such that when $\eta_i \to 0$ all observables reproduce SM values. In Fig. 1.1 that we introduced earlier, we saw a large green region where the BSM theory is perfectly consistent with all known data within the target observables of the theory,[7] and beyond the green region the theory is inconsistent with data for any number of reasons. Perhaps there is an additional state that should have been seen by the Tevatron, or perhaps the theory points there are inconsistent precision Z decay observables from LEP measurements, etc.

Now, a new experiment runs that has discovery potential. In other words, the new experiment can either confirm or exclude green regions of $\eta_1 - \eta_2$ parameter space in Fig. 1.1 after its run. This would be a transformative experiment, in the sense that parameter space that we thought before was viable is either confirmed or excluded by virtue of the experiment. Let us now suppose that after this new transformative experiment has run its course, the only points in the $\eta_1 - \eta_2$ plane that are consistent with the data are those that do not include the origin. In other words, experiment has shown that the SM is inconsistent with the data, while at the same time the BSM theory under consideration is consistent with experiment. In that case an enclosed green region of the (η_1, η_2) plane is selected, as shown in Fig. 1.5. Such an outcome would signify a BSM confirmation discovery. Note, by definition, SM falsification is a necessary byproduct of any BSM confirmation, where such a definition has the helpful additional implication that it prevents too eager researchers from conflating SM locus confirmation with BSM confirmation.

A BSM discovery would not mean we have necessarily found the unique correct theory of nature, just as prior to the transformative experiment we could not say that the SM was the uniquely correct theory. Indeed, the existence of any two or more theories that are consistent with the data is proof enough against the notion of a uniquely correct theory. Nor can we say that the uniquely true and correct theory underneath everything must be one of the ones that we have already contemplated (i.e., currently in the theory canon). In fact, it is a defensible conjecture that no theory can be complete and inviolable that emerges from finitely equipped minds and survives finitely scoped experiment, which are the twin rickety foundations for all theories.

There are numerous examples of BSM confirmation discoveries in the history of physics. One of the most celebrated early such discovery was the positron (anti-electron) posited by Dirac in 1931 [56] (with roots from 1928), which was placed in the theory canon, and then experimentally discovered by Anderson in 1932 [12]. The positron could have been discovered in earlier experimental works, such as by Skobeltsyn [45] and the Joliot-Curies [67], had they been more versed in the latest BSM prospects and ready to recognize the positron. The discovery of the positron was a BSM confirmation because the prevailing standard reference theory, and physics

---

[7]A "target observable of a theory" is an observable that the theory is designed to compute and purports to be correct.

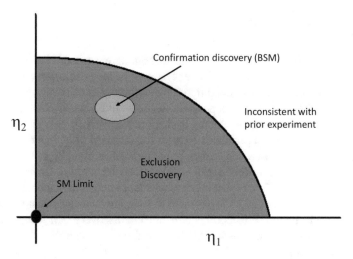

Fig. 1.5 In a BSM theory that includes parameters η_1 and η_2, and where SM predictions are obtained in the limit of $\eta_{1,2} \to 0$, one obtains a BSM confirmation discovery if future experiment rules out the origin and converges on an allowed region (green region) where $\eta_{1,2} \neq 0$

community, of the time did not agree to its necessity. Dirac was somewhat of a lone wolf crying that it needed to be there, which was strength enough to put it within the theory canon to be searched for and recognized when Anderson stumbled upon it. Other examples in more recent times of BSM confirmations are quarks/partons [30, 32], parity violation [110], and neutrino masses [65, 66, 72].

Regarding neutrino masses, it was an implicitly held view for decades that the neutrino should have zero mass and thus the SM with massless neutrinos was the default position defining the SM [74]. As Ramond puts it, "In fact neutrinos are absurdly light, to the point that it was widely believed that they were massless" [91]. It was thought that since the neutrino mass should not be so much lower than the electron mass if there existed a right-handed neutrino, the explanation for the smallness of the mass then must be the consequence of the right-handed neutrino simply not existing, thereby disallowing any pairing with the left-handed neutrino to achieve a mass term. Neutrino masses would imply the need to add to the SM either a dimension-five operator, a family of right-handed neutrinos, or more. Thus, all theories with neutrino masses were BSM theories, and began to populate the theory canon. These included theories involving a myriad of ways to naturally explain why neutrino masses are nonzero but tiny compared to other massive elementary particles in the SM [77].

Confirmation that neutrinos definitively had non-zero mass occurred in 1998 [65, 66, 72], thus excluding the standard reference theory of zero masses and consequently adjusting/redefining a new SM that incorporates neutrino masses. The discovery did not happen by accident as it required tremendous investment in state-of-the-art equipment to make the BSM confirmation discovery.

1.4 Exclusion Discoveries

Confirmation discoveries are not possible without an experiment having the capability of exclusion. The capacity of an experiment to have exclusion discovery is a necessary, albeit not sufficient, condition for a confirmation discovery. Furthermore, with a theory canon in hand, it is possible to carry out an assessment to determine if an experiment indeed can guarantee exclusion discovery. It is arguably the duty of all resource-intensive experiments to meet the standard of guaranteed exclusion discovery. As will be discussed below, pursuing such a capacity is likely only to be positive, with no inadvertent negative side effects for a priori capacity to falsify the SM or make revolutionary discoveries, which are two complementary forms of discovery that do not require an experiment to have guaranteed exclusion abilities.

There are four main sub-categories of exclusion discovery. SM locus exclusion, which seeks to exclude a locus of BSM-inspired points within the SM parameter space. SM falsification, which seeks to find evidence that the SM is inadequate to account for all experimental data within its presumed domain. BSM exclusion, which seeks to exclude regions of parameter space within a BSM theory in the theory canon. And BSM falsification, which seeks to falsify a BSM theory, by excluding its entire parameter space. Each of these will now be discussed in turn.

1.4.1 SM Locus Exclusion

In Sect. 1.3.2 above we noted that BSM-motivated considerations can lead one to predict that future experiment would narrow the experimentally allowed region within the SM parameter space to a small locus of points. We discussed how quasi-confirmation and confirmation of a SM locus could develop in the course of experimental work. It is equally of interest to note that a SM locus could be excluded upon further experimental investigation. That is the subject of this section.

Recall from Fig. 1.2 the situation of an hypothesized relationship between SM parameters c_1 and c_2 (solid black line) which is initially (at t_0) consistent with, say, the 95% CL region obtained by experimental measurements. After some time (at t_1), let us suppose that the experiment has reduced its errors significantly and is now in position to re-test whether the hypothesized relation is still viable. In this case, the 95% CL allowed region at t_1, depicted in Fig. 1.6, is the small green region, far away from the solid black line. Thus, the hypothesized relation—the locus of points that satisfy the relation—is excluded. This we call the "locus exclusion discovery."

A locus exclusion is only a meaningful discovery if it is considered part of the theory canon, meaning that the relation was of high interest to scientists for defensible reasons. In the history of particle physics there have been many interesting locus exclusion discoveries. One example was the hypothesized Veltman Higgs criterion [103], which was postulated to be satisfied to control quadratic diverges of the SM Higgs sector. The criterion states that $\mathrm{Str}\,\mathcal{M}^2(\Lambda_V) = 0$ at some scale Λ_V

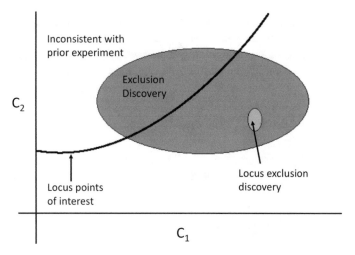

Fig. 1.6 When the SM has parameters (C_1, C_2) and a BSM theory predicts a locus points of interest (black line), a locus exclusion discovery occurs when future experiment excludes all points that lie on the locus points of interest region. This is illustrated by the experimentally allowed green region far from the locus points of interest line

where the quadratic divergences are required to cancel. Str is the super-trace (-1 for fermions and $+1$ for bosons) over masses of elementary particles in the SM. One subtlety is to know exactly what scale Λ_V one should evaluate this criterion. If one chooses $\Lambda_V = M_{\text{Pl}}$, which is the highest known putative fundamental scale and thus where quadratic diverges would be most violently destabilizing, the condition predicts the Higgs mass to be $M_h \simeq (135 \pm 2.5)$ GeV, which is now excluded by more than 3σ [53]. The SM locus of Higgs masses predicted by lower Veltman criterion scales $\Lambda_V < M_{\text{Pl}}$ are excluded by even higher significance. Thus, experiment has made a locus exclusion discovery.

An example of an extremely important locus exclusion discovery was the determination that the cosmological constant is not zero. For many years it was thought that whatever solved the cosmological constant problem probably made it zero rather than a small but non-zero number, whose scale would be very hard to justify. It was a vague notion, since quantum gravity was and still is too difficult to make such predictions, but it was a qualitative possibility that was attractive. Thus, we can say that $\Lambda_{CC} = 0$ was a locus of high interest in particle physics and cosmology. Excluding it would be a significant discovery by any definition of the word discovery. In 1998 that is what happened, when the experiments searching for an accelerating universe through supernova candles discovered evidence that the cosmological constant indeed cannot be zero, thus excluding that locus point [88, 89, 94]. This locus exclusion discovery remains one science's most significant discoveries of the last quarter century.

Finally, there is a third example that is interesting to discuss from the world of neutrino physics. In order to understand the masses and mixing hierarchies among the three generations of neutrinos one can introduce discrete flavor symmetries based on finite groups, which when neutrinos are assigned different representations under those groups give rise to characteristic mixing angle predictions. For example, a famous leading-order prediction for the neutrino mixing angles is so-called tri-bimaximal mixing [71]:

$$\sin^2 \theta_{23} = \frac{1}{2}, \quad \sin^2 \theta_{13} = 0, \quad \text{and,} \quad \sin^2 \theta_{12} = \frac{1}{3}. \quad (1.1)$$

This forms a locus point of prediction, which early neutrino data was consistent with. However, the present data [27] for neutrinos is

$$\sin^2 \theta_{23} = 0.481 - 0.570, \quad \sin^2 \theta_{13} = 0.0207 - 0.0223, \quad (1.2)$$
$$\sin^2 \theta_{12} = 0.291 - 0.318 \quad (1\sigma \text{ ranges})$$

assuming normal ordering hierarchy of neutrino masses. Thus, present data is not consistent with the locus of Eq. 1.1, and experiment has made an important locus exclusion discovery. Of course, most theories of neutrinos give rise to leading order estimates, as was stated for our prediction of Eq. 1.1. This means that the measure-zero point of the prediction is unlikely to be exactly correct, but rather only correct to within leading order, and the "true point" is in the neighborhood. One can define that more formally by declaring a locus of points that is within $\delta = 0.1$ of the values on the right-hand side of Eq. 1.1, or some other value of δ, and then assess whether there has been a locus exclusion discovery or not. In the present case, most would agree that indeed tri-bimaximal mixing has been excluded and that a true locus exclusion discovery has been achieved by neutrino experiments.

1.4.2   SM Falsification

A strong form of exclusion is the total exclusion of the SM. This is SM falsification. It is achieved by recognizing that no observable within the domain of the SM can be accommodated by any point within its parameter space. One example of how this could happen is if it were found that the Z decays into b quarks occur too often, in violation of precision SM predictions. In general, SM falsification is not a question, despite simple appearances, of a single observable deviated from the SM, since one can always choose a set of parameters to make a single observable match expectations. Rather, it is a question of whether a global analysis of all measured observables within the domain of the SM ($\sigma(WW)$, $\Gamma(Z \to e^+e^-)$, m_{top}, m_W, $A_{\text{FB}}^{t\bar{t}}$, etc.) are compatible with at least one point in its parameter space.

The only way to accomplish SM falsification, if it is possible at all, is through extensive measurements of SM-targeted observables along with extensive theoretical work that enables the comparison of experiment with SM expectations. Thus, it puts a high premium on SM analysis development within the theory community, and the growth of experimental observables pursued and the improvement of existing experimental measurements.

It is hard to imagine anybody arguing against the importance of activities that attempt to stress-test and falsify the SM. Discovery of SM falsification would be quite momentous. It is often thought that the evidence for dark matter, coming from a variety of sources such as rotation curves of galaxies and cosmic microwave background observables, is evidence that the SM is falsified already. That is perhaps a too strong declaration since the theoretical description that provides dark matter theories may be found to be complementary add-ons to the SM rather than something that rips apart the fabric of the SM and pieces it together in a new structure. That is why the SM is still considered the standard theory within high-energy particle physics despite this strong evidence that the SM cannot be complete. Thus, researchers look for other ways to falsify the SM within the presumed domain of applicability, such as high-energy interactions of its quarks, leptons, neutrinos, gauge bosons and Higgs boson.

The worthy task of falsifying the SM can at times be confused with the notion that all thinking, both experimentally and theoretically, must be purely SM-based, with no reference to any BSM notions. To those who hold strongly to this "signalism" viewpoint, BSM theories are abhorrent and should be banished from scientific discussions of high-energy physics. In Sect. 1.6 a discussion is given on the risks of attempting to pursue discovery entirely through focus on SM falsification with no reference to BSM theories. It will be concluded there that although SM falsification is very important, and SM-based theory work and experimental work is unambiguously required for progress, SM falsification is likely to be more efficiently achieved as a byproduct of the pursuit of BSM exclusion/confirmation.

1.4.3 BSM Exclusion

Consider an experiment that has the potential to make a BSM confirmation discovery. If after operating for some time a confirmation discovery does not happen, the experiment usually can place constraints on the parameter space of the BSM theory. Constraints identify exclusion regions of the theory. Although frontier experiments are hard to construct and are not commonplace, it is nevertheless common that their usual products are exclusion discoveries. Indeed, exclusion discovery generally must precede confirmation discovery.

Figure 1.7 gives a representation of exclusion discovery within a decoupling BSM theory. In the figure the SM is at the origin. As experiment gathers more data the exclusion line collapses inward, and the region between the old exclusion line and the new exclusion line is the exclusion discovery. It is to be legitimately called a discovery since there was a priori potential for the BSM theory to be confirmed within that region and there is significant new information about that region in the theory canon that was hitherto unknown. Exclusion discoveries can occur simultaneously with confirmation discoveries, as Fig. 1.5 illustrates, even though the BSM confir-

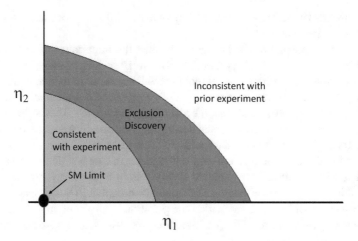

Fig. 1.7 A BSM exclusion discovery occurs when a BSM theory has a region of its parameter space eliminated (red region) by experiment, leaving behind a smaller allowed region (green region) which is connected to the SM limit point of $\eta_{1,2} \to 0$ where all predictions of the BSM theory are indistinguishable from those of the SM

mation result of that case would be the primary news trumpeted. Because of the simple necessitating role that exclusion must play in even confirmation discoveries, guaranteed capacity for exclusion discovery should be the standard by which future proposed experiments are judged. If they wish to be transformative experiments and make discoveries, they will have to extend exclusions beyond what the sum total of all prior experiments could do.

As an example, the LHC has made numerous exclusion discoveries in addition to its celebrate SM confirmation discovery of the Higgs boson. It has excluded large regions of parameter space (i.e., of the theory canon) for minimal supersymmetry and minimal models of composite Higgs. See Figs. 12 and 13 of Ref. [1] and Figs. 2–5 of [13] for schematic representations of the exclusion discoveries within these theories. It has also made exclusion discoveries for theories with leptoquarks, extra gauge symmetries, extra spatial dimensions, etc.

The frequent recitations of exclusion discoveries are at times cast as pronouncements of failures [33], but of course they are necessary and must occur frequently for progress to be assured. Instead, they should more rightly be viewed as successful executions of experimental work that had capability for such exclusion discoveries. Excluding enormous swaths of the theory canon is a major achievement of experiment, as one would quickly realize if they tried to do that at home. A vibrant and rich BSM theory canon means there will be many individual disappointments, but effort is sustained and rewarded when viewed as a collective search party exploring uncharted territory for deeper hidden knowledge.

In short, exclusion discoveries are important discoveries that are necessary and expected when experimental science progresses. The existence of an exclusion discovery signifies that a BSM confirmation discovery was possible, highlighting the experiment's value. Furthermore, well-articulated exclusion discoveries raise the bar for future experiments to have discovery potential, allowing us to question sharply experimental plans and projects that cannot demonstrate future guaranteed discovery potential.

1.4.4 BSM Falsification

There are times when experiment makes sufficiently strong exclusions that an entire BSM theory (i.e., a BSM theory's entire parameter space) is falsified out of the theory canon due to its inability to be empirically adequate. An example of this is the minimal fourth generation model, which postulates that there is another generation of fermions in addition to the three generations that we already know, and that these fermions all have degenerate (or nearly degenerate) mass. Careful precision measurements at LEP were enough to rule out this theory [6]. This can be called a BSM falsification discovery, and is an extreme version of exclusion discovery.

Another BSM falsification discovery made by Tevatron alluded to in the introduction is that of minimal no-scale supersymmetry with neutralino dark matter. In this theory, there is an upper bound on the value of $m_{1/2}$, above which the neutralino cannot be the dark matter. In a plot with decoupling parameter(s) where the SM is at the origin, this would be equivalent to eliminating by theory construction the parameter space in the neighborhood of the SM. Thus, the theory does not have a decoupling SM limit point within its definition that would give it experimental safety against continuing exclusion discoveries. At some point the exclusions run out of real estate and the theory cannot be accommodated. At this point a BSM falsification discovery has been made. A schematic representation of a BSM falsification discovery is given in Fig. 1.8, where a transformative theory has turned a once green region (i.e., consistent with experiment) entirely into a red region, such that there is no longer any experimentally allowed region for the BSM theory.

BSM falsification discoveries can be controversial especially if a theory's parameter space is reduced by non-empirical methods. For example, some might say that the LHC experiments have made a BSM falsification discovery by completely excluding supersymmetry. Another example is the purported falsification of the minimal grand unified theories, both supersymmetric [83] and non-supersymmetric [100]. However, these BSM falsification discovery require a strong cut on parameter space by non-empirical means, such as not taking into account the possibility of higher dimensional operators or taking naturalness and finetuning criteria very seriously and assuming a rather aggressive (i.e., not conservative) low cutoff value of finetuning to be acceptable [68, 104, 107, 108].

Nevertheless, if the community, or some respectable fraction of the community, deems these extra-empirical conditions to be required for (non-)admission the theory canon, then so be it. However, the BSM falsification discovery should not be declared without a very precise description of exactly what theory is being expelled from the

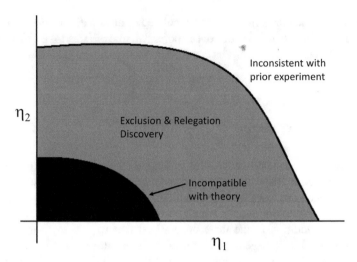

Fig. 1.8 A BSM falsification discovery occurs when a BSM theory has its entire parameter space eliminated (red region) by experiment, leaving no allowed region behind. If all observables are still consistent with the SM, a BSM falsification can only occur if the BSM theory has no SM limiting point, either by construction of the theory, or by eliminating a region around the SM limiting point of $\eta_{1,2} \to 0$ by non-empirical methods, such as by imposing limits on maximum tolerated finetuning of parameters

theory canon, including whatever additional non-empirical criteria applied. This is often lacking.

One should emphasize that although naturalness and finetuning considerations were invoked originally because it was thought that the next stage theory is likely to be natural, it has also served, perhaps subconsciously, as the only means by which a BSM falsification discovery could be made for many theories. Without it there can be no exclusion of supersymmetry, or most of its invariants, for example, according to our current understanding of the theory and the data. This is true of all decoupling theories—composite Higgs, extra dimensions, supersymmetry. Without some way to expel a finite region around the SM decoupling limit point(s), there is no way to ever falsify them when data stays consistent with SM. One just makes exclusion discoveries (or confirmation discovery) closer and closer to the SM limit point(s) in the parameter space. Naturalness also has served indirectly a very useful practical role in that it has encouraged physicists to think much more about experimentally accessible phenomena, since the decoupling region without experimental consequences is an anathema to a natural BSM theory.[8]

---

[8]For this reason, and others, it is baffling why anybody who cares deeply about theorists focusing on theories that are accessible to experiments should think it destructive to science progress that a researcher is encouraged by the naturalness criteria when theory model building. On the other hand, if theorists had become enamored with the "principle of anti-naturalness," where every new theory had to be highly finetuned for some reason, and thus typically out of reach of every conceivable experiment, that would be a significant concern to science progress. Thankfully, that never happened.

Some of the most interesting experiments are those that have made BSM falsification discoveries. "Ruling out" theories has large impact in how science progresses. Ruling out theories with the simple QCD axion, additional light neutrinos, degenerate fourth families, minimal versions of supersymmetric theories, minimal technicolor theories of electroweak symmetry breaking, minimal top-quark condensate theories, minimal $SU(5)$ grand unified theories, etc., have all set the field in different directions which were by definition more productive. Each of these BSM falsifications took prodigious experimental skill and resources to achieve, and each has had tremendous impact on high-energy physics. To increase the meaning and impact of experiments, articulating precisely all the BSM falsifications that it has achieved within the theory canon is a useful endeavor. This may involve creative categorical parsings of parameter spaces within bigger frameworks (supersymmetric, compositeness, etc.) but the results expressed are meaningful and powerful, and set the standards above which future experiments must achieve.

1.5 Revolutionary Discoveries

There are times when experiment has results that are in conflict with every theory within the theory canon. Such discoveries can be called "revolutionary discoveries" since it annihilates the entire theory canon and requires one to start anew in model building that takes into account the new results. It goes without saying that the experimental result must be solid and reproducible and beyond reproach in order to claim a revolutionary discovery.

Perhaps "annihilating" the theory canon is too strong, since the old standard theory is likely to still be of use in a domain restricted compared to what is once was before the revolutionary discovery. For example, after the muon discovery it is understand that the total cross-section of e^+e^- annihilations is higher than the old QED results for center of mass energy greater than $2m_\mu$, since $e^+e^- \to \mu^+\mu^-$ contributes now in addition to the standard $e^+e^- \to e^+e^-$, $\gamma\gamma$, etc. results of QED. However, the total rate for $e^+e^- \to e^+e^-$ is not significantly changed[9] due to the discovery of the muon and thus the old theory is still useful. Indeed, the muon can be merely added to the QED lagrangian in a way directly analogous to the electron except that its mass is higher.

Nevertheless, what is clear is that no theory as it stood before in the theory canon survives a revolutionary discovery, by definition. In the case of the muon discovery, that new theory was not terribly difficult to devise since the muon interactions were so similar to the electron interactions. However, it was a new theory, and it is reasonable to conclude that the experimental discovery was revolutionary according to standard connotations of the word.

---

[9]Nevertheless, there is a change, albeit tiny, since very precise measurements would be sensitive to quantum loops of virtual muons in the photon propagator mediating $e^+e^- \to e^+e^-$.

Some revolutionary discoveries lead to a new theory canon that is initially all incorrect. An example of this is the anomalous perihelion procession of Mercury. Once the experimental result was established by Le Verrier, it was thought that the theory needed to change or the objects that were part of the theory description (sun, planets, asteroids, etc.) needed to change. The most compelling idea was a new planet between Mercury and the Sun, but later experiment did not find it. New ideas that invoked more finely grained objects in dust belts that could not so easily be found by experiment were then invoked, but experiment ruled out parameter space more and more finely for such ideas [95]. What ultimately worked was an entirely new theory within the theory canon—General Relativity—which explained that result. It also made an additional non-trivial prediction that no other theory in the canon made, that of the bending of light. When a confirmation discovery was made for the bending of light, General Relativity sat most prominently in the theory canon for describing gravitational phenomena. In this history, the discovery of the anomalous perihelion precession of Mercury was a "revolutionary discovery" and the discovery of the bending of light was a "confirmation discovery." For those who were not convinced of general relativity yet, it was a "BSM confirmation discovery" while for others who were already convinced it was a "SM confirmation discovery." In any case, it was a discovery for all.

There are many other examples of revolutionary discoveries in the recent history of physics. Events with missing energy in nuclear β decays were unexpected and thus were revolutionary discoveries. The missing energy was ultimately explained by the existence of neutrinos, which was confirmed [84]. To some it was a BSM confirmation discovery since the discovery of neutrinos was preceded by the theoretical prediction of its existence, whereas others would view it as a revolutionary discovery since the original experimental result annihilated the entire theory canon that had existed at the time, although it took physicists some time to realize that. The discovery of the acceleration of expansion of the universe would qualify to some as a revolutionary discovery since in so many people's minds it was inconceivable that the vacuum should have a positive cosmological constant, which is what the results imply. To others it was a "locus exclusion discovery," excluding zero cosmological constant value, as was discussed above in Sect. 1.4.1.

In summary, a revolutionary discovery happens when the entire theory canon is falsified by unforeseen phenomena. Revolutionary discoveries cannot be guaranteed nor even anticipated. They happen "out of the blue." Nevertheless, revolutionary discoveries do happen, and an important remaining question of this analysis of discovery is whether focus on BSM exclusion, which has been argued to be the only path to assured discovery, dims the prospect for revolutionary discovery, which is never assured and which by definition takes place outside of the entire theory canon and its array of BSM theories. That is the topic of the next section.

1.6   Signalism: Risks of Pursuing Discovery Without BSM Context

The discussion above points to the utility of pursuing discovery with a BSM-oriented approach. Historically there is much agreement on this approach, although the language by which it is phrased may be different. For example, as A. P. Aleksandrov reports, "[Euler] himself believed that science progresses via conjectures, by successively rejecting less accurate conjectures in favor of more complete ones" [31]. And Feynman said,

> In general we look for a new law by the following process. First we guess it. Then we compute the consequences of the guess to see what would be implied if this law that we guessed is right. Then we compare the result of the computation to nature, with experiment or experience, compare it directly with observation, to see if it works. If it disagrees with experiment it is wrong. In that simple statement is the key to science. [62]

These two quotes clearly highlight the importance of the BSM confirmation/exclusion approach to discovery, in our language.

Nevertheless, there creeps in a counter-sentiment to the BSM confirmation/ exclusion approach to discovery, which Feynman also brings up in the same lectures wherein the above quote was delivered:

> This [the BSM confirmation/exclusion approach, in our language] will give you a somewhat wrong impression of science. It suggests that we keep on guessing possibilities and comparing them with experiment, and this is to put experiment into a rather weak position. In fact experimenters have a certain individual character. They like to do experiments even if nobody has guessed yet, and they very often do their experiments in a region in which people know the theorist has not made any guesses. For instance, we may know a great many laws, but do not know whether they really work at high energy, because it is just a good guess that they work at high energy. Experimenters have tried experiments at higher energy, and in fact every once in a while experiment produces trouble; that is, it produces a discovery that one of the things we thought right is wrong. In this way experiment can produce unexpected results [revolutionary discoveries in our language], and that starts us guessing again. [62]

From these two quotes of Feynman above, within the same lecture we see two different approaches to pursuit of discovery: the BSM confirmation/exclusion approach and the no previous "guesses" approach to experiment.[10] Feynman does not resolve that tension in his lecture, or comment further on the relative merits of each. Perhaps a healthy balance of both approaches is the best path, one might wish to think. It will be argued here that a primary orientation to BSM exclusion/confirmation is superior, while at the same time it does not diminish the prospects for surprises. Seeking pure surprises outside the context of a BSM orientation is a riskier endeavor.

To develop and define the tension further, let us first acknowledge that, as with Feynman's "experimenters" in the previous quote, there has been and is now a current within particle physics that is reluctant to embrace theoretical speculations except in times of crisis when no theory is empirically adequate at all. For example, when

---

[10]One is tempted to call this latter approach the "shut up and build" approach to experimental science.

β-decay was shown to have what looked to be violations of energy conservation, speculation as to what could cause that signal was acceptable. When new signatures were found in cosmic rays, which ultimately led to our conceptualization and discovery of the muon, speculation as to what the origin was of that signature was acceptable. In contrast, speculations that are not in the service of an extreme experimental crisis are considered by some to be "philosophy" that has very low efficiency in revealing truth. To such individuals, it is much better to become divorced entirely from BSM consideration and approach experimental studies in a "model independent" way or in a "signal-based way." The desire, in other words, is to seek SM falsification or revolutionary discoveries with no reference or consideration of BSM rationales. A key assumption of this mindset is that focus on BSM exclusion/confirmation discoveries derails experiment from the more productive focus on SM falsification or revolutionary discovery pursuits, which then as a byproduct creates a crisis—the standard theory being unambiguously incompatible with experiment—and thus opens the door to focused, fruitful, BSM theory. In the absence of such a crisis there should be no BSM theory work; there should be only SM theory work so as to more effectively identify when the SM has been falsified and when a true revolutionary discovery has indeed been made. This approach can be called "signalism" with its emphasis on initiating and studying signals, then comparing them with SM expectations, all at the exclusion of any BSM theory context.

Again, it will be argued in this section that signalism poses a significant risk to discovery of every kind, including its targets of SM falsification and revolutionary discoveries. One of the themes that will be explored below is that no matter how hard one tries, escaping BSM considerations is impossible if one wants to give meaning and context to the SM and have any rational basis for future experiment. As we will discuss, to make an argument for a future experiment or experimental analysis is necessarily to engage in BSM theory assessment, whether one recognizes it or not.[11]

Let us start by giving some of the arguments for and against the proposition that BSM-exclusion/confirmation approach to planning and executing experiment reduces prospects for revolutionary experimental discoveries. On one side of the argument, one could claim that focus on BSM exclusions within the theory canon promotes a more narrow set of experiments aimed at the narrow set of phenomena that the BSM canon theories predict, leaving out searches for the vastly greater array of new phenomena that a less theory-laden approach to experiment could probe. On the other side of the argument, one could claim that the vast majority of discoveries of the last century are BSM confirmations and not revolutionary discoveries, and that with finite resources (i.e., we cannot cover/measure all conceivable phenomena one could think of anyway) the search for BSM exclusion/confirmation is the best investment. In any event, focus on BSM exclusion/confirmation is an experimental activity that a priori has arguably at least just as much chance of finding something unexpected (i.e., revolutionary) as any other conceivable approach.

---

[11] Analogs to this "you cannot escape speculative theory" argument can be found everywhere in intellectual pursuits, as far and wide even as literary theory: "Hostility to theory usually means an opposition to other people's theories and an oblivion of one's own" [58].

Furthermore, even if one did not like BSM-exclusion/confirmation centered approach to experimental searches and wanted to instead focus on initiating a program to maximize revolutionary discoveries, a curious contradiction develops. The moment one seeks a rational description of an investment with aims toward revolutionary discovery is the moment one unwittingly entertains nebulous ill-conceived BSM notions well outside what experts in BSM theory would think is interesting or that could possibly solve any recognized problem that the SM does not address. And as soon as an argument ensues that the established BSM experts are wrong in their assessments, and that one's vision of what is possible in BSM physics is defended, the researcher then becomes a BSM theorist, losing their revolutionary-only claim. Thus, attempts to seek only revolutionary discoveries in research, while rebuffing any and all BSM notions, necessarily dissolves into the mystical and visionary and away from the rational. In such a universe of thought that consciously runs away from rationales it is hard to decide whether throwing a vase off the Eiffel Tower to see what new thing might happen is better than analyzing high-luminosity LHC data.

To further explore these issues, one approach to try to be a revolutionary experimentalist is to decide, consciously or subconsciously, that any BSM phenomena or signal (not necessarily derived from a theory) ever expressed that is incompatible with the SM but not in contradiction with prior experiment is currently viable, and experiment could be chosen to pursue a randomly chosen (because of limited resources) new signals identified among them, thereby eliminating "theory bias." However, this approach is yet another BSM theory, which requires the researcher to believe that a randomly chosen new phenomenon/signal identified by humans that is not possible within the SM, but which is not yet excluded by experiment, is more likely to be manifested by nature than any phenomena derived from theories constructed for the purpose of solving outstanding problems in the SM. There is no known logical justification for such a signalist belief.

A closely related line of argument would be to claim that pure focus on BSM exclusion/confirmation necessarily dims the prospects of revolutionary discoveries since it prohibits signalism-like approaches to science from being funded and pursued. However, the missing element of that argument is that signalism has yet to be justified, and once it is justified sufficiently to garner resources it becomes an established BSM theory, and we are back to focus on BSM exclusion/confirmation. This is not to mention that it is hard to imagine any compelling justification for signalism, or related ideas, since the number of very odd signals possible versus those realized in nature is arguably infinite and thus the probability would be vanishingly small for selecting a good one to pursue in the search for a revolutionary discovery. This leaves standard (non-signalismic) BSM exclusion/confirmation as a preferred focus for discovery.

An example illustrating this is in the career of Nobel Prize winner Martin Perl, whose attitudes and public pronouncements can best be described as signalismically oriented, but whose activities mixed BSM exclusion/confirmation with signalism-based approaches, with its inevasible slide toward BSM justifications (see, e.g., [78]). Perl frequently made the statement that experimentalists should be cautious about, and perhaps even ignore, popular physics theories (i.e., the BSM theory canon), as evidenced by his 1986 essay [87]. In that essay, he made several claims that are at odds with the thesis of this essay. For example, Perl wrote,

> Experiments based on speculative theories and with narrow goals teach us little if the answer is no — only that the theory is wrong or, more likely, that the parameters in the theory need adjustment. [87]

However, there are two problems with this statement. First, he conflates "experiments based on speculative theories" and "experiments with narrow goals." This is a false dichotomy on experiment. Experiments can be extraordinarily ambitious and broad and yet target "speculative theories" (BSM theories).

Furthermore, Perl makes the claim, in our language, that BSM exclusion discoveries "teach us little." As opposed to Perl, we have argued here in this essay that if the SM is not falsified and a BSM confirmation/exclusion discovery has not occurred, then we have learned very little. The primary knowledge that most experiments in history have taught us is through BSM exclusion/confirmation discoveries. It teaches us in all the ways discussed above, including creating a new threshold above which future experiments must achieve to be deemed worthy successors. In Perl's essay there is not a developed argument for what it means for experiment to teach us something, but one might infer that, in the absence of SM falsification and BSM confirmation, what it can teach us primarily in Perl's view is the recital of unadulterated experimental data, to be written on velum, metaphorically, and stored for humanity's gaze in perpetuity. However, there is nothing incompatible with this majestic view of experimental data having worth in and of itself, and the additional view that it can be utilized to monitor and characterize the status of BSM canon theories. After all, it should never be lost that the SM is not the only theory compatible with all data. Ironically, only a deeply held non-empirical mindset, which necessitates strong speculation, would demand intense loyalty to the SM only, and show disdain for competing empirically adequate theories in the canon.

Another closely connected question is whether focus on BSM-exclusion/ confirmation dims prospects of falsifying the SM. This is especially problematic for those who adhere to the signalism viewpoint. According to the signalism mindset, the primary path of science is to decide on the Standard Theory, which in the case of high-energy physics is the SM, and once it is decided, one should focus entirely on falsifying it. In other words, the only activities of any value are those that stress-test the SM to the extreme with the hopes of breaking the SM, either through unambiguous statistical anomalies of SM observables or through revolutionary discoveries, such as new non-SM resonances. Only after the SM is broken should we pick up the pieces and do fancy theorizing that constructs a new SM. Any activities that build and analyze more ambitious BSM theories are terrible wastes of time and very inefficient, since more or less all ideas except at most one must necessarily ultimately crumble to dust with more knowledge, which by the way is not guaranteed anyway. For this reason, again, the only theory work should be that which computes a myriad of experimental observables to higher and higher accuracy with an aim to reducing the allowed SM parameter space to the smallest possible volume.

It must be acknowledged that the pursuit of SM falsification is unquestionable a coveted achievement in high-energy physics since falsification would imply that qualitative new understanding of nature is needed to establish a new SM that accommodates the data. However, narrow focus only on SM exclusion without the benefit of a BSM perspective risks derailing the very falsification goals it is trying to achieve. For example, a narrow SM-only focus does not imagine any ways that observables could go awry and only attempts to get the very best measurements possible to squeeze the allowed parameter volume of the SM compatible with experiment to smaller and smaller values.

For example, a SM-only perspective could suggest that instead of venturing into an energy frontier one could only increase the intensity frontier, gathering more and more Z boson decays, and more and more W's and top quarks to compare with precision measurements. Measuring high-energy e^+e^- or pp collisions at significantly higher energy but with somewhat limited luminosity appears worthless to a SM-only perspective. Without a BSM perspective it is hard to ever imagine a strong reasoned case for going to the high energy frontier. No parameters of the SM will be measured better by such new endeavors in many cases. Yet, we know that from the perspective of the SM being an effective theory of a more extensive higher energy theory that the effects of new physics, which if seen would break the SM, often become more and more pronounced at higher energies. The development of SM-incompatible signals (SM falsification) at higher energies is due to momentum-dependence of higher dimensional operators in the effective theory, or the opening up of new particle thresholds. This is decidedly a BSM-perspective, developed by vast experience with BSM theories, and could never be divined from an unadulterated SM mindset.

In contrast to a BSM-informed mindset, the pure SM mindset could keep one anchored forever to pursuing high statistics right around the weak scale in order to measure SM parameters better and better, with hope that one day precision statistical analysis would develop a deviation that grows over time, not even considering the possibility that a new particle or new interaction could be not far away in the energy frontier or in the frontier of new experimental methods. Again, as soon as one even contemplates a motivation to spend more money and go to higher energy because, for example, "a new particle might be there," they have entered the BSM world whether they like it or not, and thus must face the fact that their speculative motivations to spend more money and build a costly energy frontier facility could be challenged or criticized, and thus must be justified in some way at least as plausible. In other words, they need to articulate why their BSM speculations are worth pursuing. A vibrant community of BSM scholars can then be appealed to for that task, and thus helpful in the informed co-pursuit of SM falsification, even when trying to falsify the SM is the experimentalist's only goal.

Arguments of discovery centered on BSM exclusion/confirmation vs. signalism abound in less august forums, such as social media posts, blogs, and letters to the editor pages of Physics Today. A particularly prominent one from the early 2000's was Harry Lipkin's *crie du coeur*:

> I have no patience with social scientists, historians, and philosophers who insist that the 'scientific method' is doing experiments to check somebody's theory. The best physics I have known was done by experimenters who ignored theorists completely and used their own intuitions to explore new domains where no one had looked before. No theorists had told them where and how to look. [79]

and Lincoln Wolfenstein's equally forceful retort that lists theory confirmation after theory confirmation discovery and ends with

> We do not have a theory of everything, although some of my colleagues dream of one. When new domains of energy are explored, we will not be surprised to discover that there are things in the heavens and on Earth that are not described by our present theory. Our goal, then, must be to find a more encompassing theory and design experiments to fully test it. That, I believe, is the scientific method. [109]

In the end Lipkin's signalism-oriented viewpoint, springing from a rather cursory historical analysis, emerges naive compared to Wolfenstein's BSM-oriented view. That the most consequential physics discoveries have "ignored theorists completely," as Lipkin claimed, is rather easily countered with the simple realization, for example, that no experiment ever measured the process $e^+e^- \to \mathrm{SU}(2)_L$.

In summary, signalism is the desire to make SM falsification discovery or revolutionary discovery through construction of signal-based analyses that compare experiment with SM theory, while disallowing all reference to speculative BSM theories. The problem, as discussed above, is that tacking to revolutionary discoveries without BSM reference is an inscrutable exercise that can never have a rational justification.[12] At the same time, tacking to SM falsification discovery without BSM reference can lead to experiment pirouetting longer and longer on the same or similar experimental analyses, accruing more and more data over time as they improve SM parameter determinations with hopes of a statistical incompatibility developing, meanwhile feeling very little pressure to think of and pursue different types of experiments that test more fundamental theory lying in wait "just beyond." That is not to say that achieving higher precision on SM observables is not worthwhile physics. However, when deciding which parameter to pursue to much higher precision, BSM insight helps decide. BSM also helps decide when results of a particular experiment and signal are good enough, and something else should be done. Focus on BSM physics is key for pruning away inscrutable or unproductive pursuits, for guaranteeing discovery (BSM exclusion/confirmation), and for enhancing prospects of SM falsification and revolutionary discoveries.

---

[12]The risks of pursuing revolutionary discoveries through new experiments without any theory context allowed, which then does not allow comparisons of value with respect to prior experiments and observations, has been illustrated well recently by Caldwell and Dvali in the specific case of anti-matter gravity experiments [37].

1.7 Gravity Waves and Higgs Boson Discoveries Through the BSM Lens

Two of the most momentous discoveries in particle physics and cosmology in the last decade have been the Higgs boson and gravity waves. In this section these two discoveries are compared and it will be argued that the excitement for the discoveries comes not because of the rush that comes from finally seeing and confirming a standard theory feature that we have been talking about as a science community for such a very long time. Although that is of high interest, that is not the core reason why the community has demonstrated so much enthusiasm over these discoveries. Rather, the intense excitement is to be understood as the recognition that a new era has been ushered in of present and future BSM exclusion/confirmation discoveries that were heretofore inconceivable.

Regarding gravity wave physics, let us make a few remarks about our venturing away here from traditional particle physics to discuss it. Up until now we have mostly focused on experiments very closely tied to the SM and its BSM extension, which has lead us primarily to discuss experiments that have reproducible human-induced conditions and phenomena which then is measured by experiment. However, the discovery conceptualization discussed above and the central role that BSM is argued to have are applicable to any forefront, basic science field. This is especially applicable to cosmology which shares its intellectual domain with high-energy physics for the simple fact that early time (cosmology) means high energy (high-energy physics). Gravity wave physics has breadth across many early epochs of physics, including early universe cosmology. There are standard reference models for star formation, and standard reference models for binary merger theory, and standard reference models for cosmological evolution (radiation domination then matter domination), etc. All of these can be framed as theory canons with experiments that make exclusion/confirmation discoveries with respect to them.

The discovery of gravity waves came with much fanfare, even though they were completely expected. The last reasonable scientist to be unsure if gravity waves really existed was Einstein in 1936 [75]. Since then there has been no serious questioning of their existence within the physics community. Thus, one might be tempted to say that the discovery of gravity waves by LIGO experiment was merely a boring SM feature confirmation discovery and did not advance science more than what we already knew with very high confidence.

So what exactly is it that is so exciting about the LIGO detection of gravity waves if it was a merely confirmation of what we already were sure of? The answer is that it opens the door to a vast array of guaranteed BSM exclusion/confirmation discoveries in the future. It is the reason why the "SM discovery" of gravity waves did not end the field, but rather marked the start of a new discovery era of BSM exclusion/confirmation discoveries. We now have a better idea of what the rate of the gravity waves are and we have reached the experimental threshold where we can actually measure them and plot their waveforms. Again, it is not the discovery of gravity waves themselves that is so exciting, but its implication for future discovery.

Already, there are innumerable studies of BSM ideas for early universe cosmology and astrophysics that will be tested and excluded/confirmed, leading to a vastly deeper understanding of nature. These ideas of testing new physics [36] are wide-ranging, including probes of first-order phase transitions [24], tests of early universe equation of state [48, 49, 93], probes of axions physics [90], tests of and early kination phase in the early universe, and much more, all made possible by the discovery of gravity waves.

In contrast, the Higgs boson discovery came with great fanfare for two reasons. First, unlike gravitational waves, the Higgs boson was considered by many to be a speculative possibility even moments before its discovery was announced in 2012 [106]. Thus, the discovery in and of itself was much more significant to science than the discovery of gravity waves in and of themselves.

Equally important, and more directly analogous to the gravity wave discovery, the discovery of the Higgs boson heralded the dawn of a new era of BSM exclusion/confirmation discovery. Careful measure of the Higgs boson mass enables tests of the possible composite nature of the Higgs boson [38]. It can test alternative forms of the electroweak symmetry breaking potential, including $|H|^6$ terms that could enable a first order phase transition [70]. It opens a portal to hidden worlds made possible by the only dimension-two operator in the SM that is gauge invariant and Lorentz invariant ($|H|^2$) [86, 97]. It can test supersymmetric theories since Higgs mass is computable from the supersymmetric spectrum [57]. It enabled tests of cosmological ideas, including stability of the universe [53] and even inflationary theories [28]. Precision Higgs boson production and decay measurements will enable innumerable exclusion and confirmation discoveries of BSM ideas through SM locus confirmations (e.g., predicted ratios of Higgs branching ratios) and BSM exclusion/confirmations (e.g., exotic decays of Higgs boson), all of which were made possible for the first time by reaching the experimental sophistication of copiously producing and studying the Higgs boson [50, 52]. This richness of BSM exclusion/confirmation discoveries made possible by the discovery of the Higgs boson forms the center of the physics cases for many new facilities, most particularly the ILC [64] and CLIC [10] which have stages that are designed exclusively for these opportunities.

There are other SM feature discoveries that are being made continuously that have not garnered as much attention. For example, recently LHCb has discovered two new resonances, $\Sigma_b(6097)^+$ (buu boundstate) and $\Sigma_b(6097)^-$ (bdd boundstate) [4]. There are any number of reasons why such a discovery is of high value within particle physics. Above all, it is learning more about nature—the discovery and confirmation of the kinds of structures that are allowed. It enables additional data to aid development of computational techniques. It might one day be useful in testing new physics ideas in ways that are not currently anticipated. Nevertheless, it does not have the same panache as hadronic resonance discoveries of years ago.[13]

---

[13] As Martin Perl put it, "20 years ago the discovery of an additional hadronic resonance was an important event in our world; now such a discovery gains no recognition beyond a new entry in the particle data tables" [87].

Why are discoveries of new hadronic resonances not met with tremendous fanfare in the science community? The reason is that they no longer are thought to have much impact on BSM exclusion/confirmation discoveries today. When the J/ψ was discovered in 1974, and the $\Upsilon(1S)$ in 1977, there was significant excitement due to what could be considered a BSM confirmation, in the sense that the standard theory was not settled on the necessary existence of charm quarks or bottom quarks until the discoveries were made.

In conclusion, it is not the nature of the SM feature confirmation itself that makes its discovery highly momentous in the progress of high-energy physics. Depending on circumstances a new hadronic resonance can signify a revolution ("November Revolution" of 1974 [26] with J/ψ discovery) or something far short of that ($\Sigma_b(6097)^{\pm}$ above). Rather, the significance of a SM feature confirmation lies in the impact that the discovery has on BSM exclusion/confirmation immediately upon its discovery (such as Higgs boson and J/ψ) and the significant opportunities opened up for future BSM exclusion/confirmations through new experimental portals (such as through precision Higgs boson and gravitational wave studies) recently made possible, which test BSM theories in ways previously inconceivable.

1.8   European Strategy Update

In this section we wish to add further justification to the articulation of various categories of discovery (exclusion, confirmation, revolutionary) as presented above, and evidence that all these forms of discovery are important to physicists. To accomplish the first we must show that particle physicists indeed use language that either mimics or evokes the categories described. To accomplish the latter, we must present evidence that physicists are willing to spend resources to bring about any of the various kinds of discovery. To a non-physicist the resources are monetary expenditures, but to the physicist it is their time, which is used to think and build, along with raising funds that enable them to think and build.

An excellent case study in which to investigate these questions is the European Strategy Updates. These occur in five-year intervals and are meant to set the agenda for European high-energy physics until the next update. This agenda is, at its core, a question of how resources are to be spent. This is enlightened, of course, by the science that physicists would like to accomplish. European budgets for particle physics, and CERN in particular, are relatively stable over time, and so the discussions are not about whether or not science should be done, but rather they are about exactly what science is to be done. This puts a healthy primary focus on what activities and pursuits scientists find most valuable, subject to reasonably well-known financial constraints.

We are presently in the midst of the 2018–2020 Update for European Strategy. A call for input to the process was made in early 2018 with a request that all written material to be submitted to a central repository by end of 2018. The submitted documents are then to be taken into account by the CERN council and other stakeholders in the planning of European high-energy physics projects. Deliberations and town hall meetings will occur during 2019 and a final report will be submitted in 2020. More details on the process can be found at its central website [60].

1.8.1 Framing the Strategy Update

To obtain an initial understanding of the community's conceptualization of worthwhile discovery, we look at the introductory document from the European Strategy committee, wherein they attempt to seed the discussion by saying (see "about" tab linked at [60]),

> Understanding the properties of the Higgs boson (which was discovered at CERN just before the previous strategy update) remains a key focus of analysis at the LHC and future colliders, as are precision measurements of other SM parameters and searches for new physics beyond the SM.

In our language the first two activities (understanding Higgs properties and precision measurements of other SM parameters) has discovery value in locus confirmations or exclusions,[14] and also in the attempt to break the Standard Model, or in other words, make a SM falsification discovery. Increasing precision may cause increasing tension between observables that have to obey the correlative predictions of the Standard Model theory, and when that tension becomes too great the theory can no longer describe the data and is therefore falsified. It is not necessary that the precision measurements that falsified the SM be consistent with another theory in the theory canon for a discovery to be made, of course. All that was necessary to declare a discovery is that the SM was relegated to the dustbin. We should note that these two activities, as stated, do not guarantee discovery. They merely give the prospect for a SM falsification discovery. It would be a stretch and indignity to the usual usage of "discovery" to call more and more precise measurements of the SM parameters to be an "exclusion discovery."

One should pause to discuss again the asymmetric way in which the SM is viewed compared to other empirically adequate (BSM) theories, such as minimal supersymmetry with heavy-enough superpartners to have escaped detection, or the SMEFT, which has additional non-renormalizable operators in the theory beyond the SM's renormalizable ones. In the non-SM theories, experiments that exclude regions of parameter space are making exclusion discoveries, whereas excluding some regions of parameter space within the SM (such as narrowing the experimentally allowed top mass) is not considered here to be an exclusion discovery. This distinction reflects the sentiment of the community, which holds that the SM is a special reference theory that is the "default correct" theory. Discoveries can only be made, if we are to use the word "discovery" in a meaningful way, by either falsifying the SM, i.e. showing that no point in parameter space accommodates experimental measurements, or by

---

[14]For example, continued precision measurements of the top quark mass and the Higgs boson mass to determine if, under some simple assumptions, the universe is metastable [53].

reducing the parameter space of a non-SM theory, radically through confirmation or less radically by small exclusion improvements.

The third activity listed in the European Strategy statement above is "searches for new physics beyond the SM." This is a clear call for the pursuit of either revolutionary discovery of something not thought of by physicists before or through confirmation discovery with respect to an empirically adequate non-SM theory that resides presently within the theory canon. Revolutionary discovery would be great, but it is not possible to discuss it rigorously except to say that it would be interesting if something showed up that we have never thought about before. As we will see below that is the reason that with respect to searches for new physics beyond the SM, the discussion primarily centers on how a new experimental project may be able to make exclusion or confirmation discoveries within the theory canon.

1.8.2   Discovery at Colliders

Many studies have been initiated and completed with the goal of contributing to the European strategy update. Contributions range from motivated small table-top experiments to next generation colliders on the energy frontier. In this subsection we will take a look at some of those contributions and ask in which ways their underlying conceptualization of discovery matches and differs from what has been described above. In particular, we look at the physics motivations for upgraded and new colliders. An excellent general argument for the utility of colliders is provided in [69]. What follows below is an example analysis of specific discovery goals, and how they are characterized, at the high-luminosity and high-energy upgrades of the LHC (HL-LHC and HE-LHC, respectively).

BSM at HL/HE-LHC

Perhaps the document that puts BSM physics most transparently at the center of its discussion is a report from working group 3 on "Physics of the HL-LHC, and Perspectives at the HE-LHC" entitled "Beyond the Standard Model Physics at the HL-LHC and HE-LHC" [44]. The introduction strongly indicates that discoveries in the true sense of the word have already taken place at the LHC:

> The lack of indications for the presence of NP [New Physics, i.e. BSM physics] so far may imply that either NP is not where we expect it, or that it is elusive. The first case should not be seen as a negative result. Indeed the theoretical and phenomenological arguments suggesting NP close to the electroweak (EW) scale are so compelling, that a null result should be considered itself as a great discovery. [44]

Here the authors have taken a slightly different view of discovery than has been advocated above. They implicitly require that for a BSM exclusion to be labeled a discovery it requires that the new physics be "expected" at the facility, which is a stronger requirement than our conceptualization of BSM exclusion discovery, which is perhaps best stated as an exclusion of parameter space that is "not un-expected," which is an important distinction.

Now, it is not the intention in the above paragraph to declare that expectations for preferred parameter space regions of BSM theories are not meaningful. Rather, the intention is promote the label of discovery to BSM exclusions in parameter space regions that presently may not be considered *terra prima*, since assessments of coveted lands versus non-coveted lands can change overnight for any number of reasons in physics, just as it can in geography (i.e., worthless arid lands may strike oil beneath). It should be noted that these views are not inconsistent with the report, but it is worthwhile explicitly making this point.

The BSM report continues

> A crucial ingredient to allow a comparison of proposed future machines is the assessment of our understanding of physics at the end of the HL-LHC program. Knowing which scenarios remain open at the end of the approved HL-LHC allows one to set standard benchmarks for all the interesting phenomena to study, that could be used to infer the potential of different future machines. [44]

In other words, it is the duty ("crucial ingredient") of anyone advocating for a new facility ("proposed future machines") to fully assess what prior experiment has done with respect to exclusions (what doesn't "remain open") in the BSM theory canon ("scenarios"), and future facilities must have BSM exclusion capacity beyond that ("infer the potential"). This is fully in accord with our conceptualization of discoveries and the threshold of demonstrated scientific capabilities required of future machines.

The BSM report is almost exclusively devoted to the analysis and prospects of specific BSM scenarios where BSM confirmation/exclusion discoveries can be made. However, one section is devoted to "signature based analyses", where it is introduced with

> Several contributions that are constructed around experimental signatures rather than specific theoretical models are presented in this section. This includes analyses of dijets, diphotons, dibosons and ditops final state events at HL- and HE-LHC.

This is an acknowledgement that some who are interested in making discoveries may have "signalistic" tendencies (see Sect. 1.6), who are uncomfortable thinking directly about BSM theories and hope that revolutionary BSM discoveries can be made by analysing signatures (i.e., categories of events) that are manifestly incompatible with SM expectations. However, the irony of this section is that every one of these makes reference to BSM scenarios, illustrating our claim that one cannot escape BSM even if one tries. In the list below the titles of each subsection are given along with the BSM physics explicitly invoked in the analysis, which of course goes well beyond just defining a signature:

- "Coloured resonance signals at the HL- and HE-LHC": introduces BSM diquarks into the spectrum.
- "Precision searches in dijets at the HL- and HE-LHC": introduces a BSM "new resonant state decaying to partons".
- "Dissecting heavy diphoton resonances at HL- and HE-LHC": introduces "a heavy resonance X, which decays via other new on-shell particles n into multi- (i.e., three or more) photon final states."

- "Prospects for diboson resonances at the HL and HE-LHC": introduces new "resonances decaying to diboson (WW or WZ, collectively called VV where $V = W$ or Z) in the semileptonic channel where one W-boson decays leptonically and the other W or Z-boson decays to quarks ($\ell\nu qq$ channel)."
- "Prospects for boosted object tagging with timing layers at HL-LHC": introduces jets that obtain a mysterious BSM "boost ... by $v = 0.98$", which less mysteriously could be produced "from Z decay, where the Z has been produced in the decay of a 1 TeV diboson resonance."
- "High mass resonance searches at HE-LHC using hadronic final states": introduces BSM "new resonant states decaying to two highly boosted particles decaying hadronically."
- "On the power (spectrum) of HL/HE-LHC": introduces BSM resonances that "show up in Fourier space, after performing a Fourier transform on the relevant collider data."

One is tempted to conclude that "signature based analyses" are really BSM exclusion/confirmation searches for theories that are submissively recognized to not be prominent members the BSM theory canon, which it is perhaps hoped releases the practitioners from the responsibility of detailing them and defending them as legitimate theory targets of analysis. Although there might be an element of that at times, the impetus is often to conduct an analysis that is relevant to many theories, with less close ties to any particular model.

However, a BSM-centered focus has a different perspective to "signal based analyses." It is the BSM-specific theories that should be of interest primarily. BSM centered work recognizes that a BSM theory leads to many signatures that experiment should cover, even signatures that we have not thought of before. A signal-based approach implicitly advocates the opposite direction of work: a particular signal can originate from many theories, even those theories you have never thought of before. Which approach is best? The BSM centered approach springs from solving *physics problems* (dark matter, baryogenesis, hierarchy, unification, flavor, etc.), whereas the signature-centered approach, when taken very seriously, springs from "doing" something new with no other rationale. The signature-based approach is a slippery slope toward throwing vases off the Eiffel Tower to see what happens, as discussed in Sect. 1.6. A detailed analysis of a BSM theory, on the other hand, puts pressure on defending the theory canon from whence it sprang, and it in no way impedes discovering other "new physics scenarios" that are in the same signature class.

A different strategy to re-center the "signal based" discussion of the report toward a BSM perspective, which would alter not just the presentation but also perhaps the work itself in places, would be to declare that these theories under discussion might be a bit odd, and that one cannot necessarily have much confidence perhaps that nature has chosen them, but they are empirically adequate, and each of them does satisfy at least some expectation criteria for new physics. Furthermore, they do give very different signatures compared to other theories under consideration, and perhaps would otherwise be missed if we did not take them into account. Thus, they should be admitted to the theory canon for further analysis. Now, this type of argument is indeed a reasonable argument for admitting the theory to the BSM theory canon based on 'diversity' of signatures predicted and theoretical humility [104]. One should make that case explicitly. See if people will agree when all the information is laid out. They will agree if a decent case is made. There is no need to crouch behind pseudo-"signal based" categories.

SM at HL/HE-LHC

The "SM community" of researchers have also released a report: "Standard Model Physics at the HL-LHC and HE-LHC" [21]. The main purpose of the document is to "summarise the physics reach of the HL-LHC [and HE-LHC] in the realm of strong and electroweak interactions and top quark physics...." [21]. Furthermore, it is projected that a central task is to attempt to falsify the SM by the accrual of significant amounts of data to enable percent-level precision comparisons between theory and experiment. The reasons to do this are summed up in their introduction:

> In addition, a considerable improvement is expected in precise measurements of properties of the Higgs boson, e.g. couplings measurements at the percent level, and of Standard Model (SM) production processes.... Anomalies in precision measurements in the SM sector can become significant when experimental measurements and theoretical predictions reach the percent level of precision, and when probing unprecedented energy scales in the multi-TeV regime. These anomalies could give insights to new physics effects from higher energy scales. [21]

As we have discussed above, the desire to obtain higher and higher precision of observables is an obvious activity to stress-test and possibly falsify the SM. However, the SM parameters are already measured nearly as well as they will ever be measured at the LHC, even with higher luminosity. If we look at all the parameters of the SM, including the gauge couplings, Yukawa couplings, and the Higgs self coupling, there are only minor improvements to be had by HL-LHC.

For example, the current top quark mass determination at ATLAS is $m_t = 172.69 \pm 0.48$ GeV [2] and at CMS $m_t = 172.44 \pm 0.48$ GeV [76]. Preliminary CMS projects conclude that with $3 \, \text{ab}^{-1}$ the 14 TeV HL-LHC can halve this uncertainty [21]. Likewise, the W mass currently has an uncertainty of 12 MeV [102]. Projections of uncertainty projections for the W mass at HL-LHC are approximately 50% improvement. Although this is clearly progress, it is not possible to assess what level of progress or significance this improvement has without reference to BSM physics.

The significance of precision W boson measurement, electroweak vector boson production rates and differential cross-sections at higher energies are especially dependent on BSM context. Within a pure SM mindset, such measurements have no special motivation or value if they do not increase the precision of any SM parameter above what was done long ago at LEP. To know what measuring W^+W^- scattering at the highest possible energy is more valuable than improved measurements of the Λ_b mass—neither one of which improves knowledge of the SM itself at all—comes only from recognizing that BSM theories are more likely to insert themselves and disrupt the SM expectations of the high-energy $\sigma(W^+W^-)$ measurement than the m_{Λ_b} measurement.

It is for these reasons discussed above that the "SM at LHC" document contains so much discussion on BSM physics. There is discussion of four-fermion operators, pseudoscalar colour-octets, exotic top quark dipole moments, exotic non-SM quartic gauge couplings, exotic diphoton and dilepton resonances, etc. The discussion signifies and implicitly declares that SM analyses (experimental and theoretical) can have no interesting justification and no profound value without a BSM context.

1.8.3 Discovery Beyond Colliders

In addition to projects that involve colliders, there are many interesting projects that impact the high-energy physics frontier that do not involve colliding particles at the highest possible energies. This is abundantly clear in the neutrino physics programs around the world, which focus on high-precision measurements by detectors sensitive to various species of neutrinos at various energies. Astrophysical measurements aimed at detecting non-standard deviations in cosmic microwave radiation, cosmic strings, axions and dark matter are also part of this scientific endeavor.

One recent document submitted for consideration during the European Strategy Update is the "Summary Report of Physics Beyond Colliders at CERN" [11]. Many projects and ideas are presented. An implicit recurring theme is that for a project to be justified it must have BSM exclusion/confirmation discovery capability that extends beyond any other experiment of the past, or of other experiments (including colliders) approved on the horizon.

The focus, therefore, is on BSM exclusion/confirmation of theories that are very weakly coupled to the SM ("dark sectors") and especially those that are of small mass and thus buried in the background of the more traditional collider experiments. New, lower-energy experiments are needed with specialized capacity to make these kinds of BSM discoveries. Examples of BSM targets are "are dark matter, messenger particles to dark matter, explanations of the $(g - 2)_\mu$ anomaly, the proton radius anomaly, stellar cooling anomalies and many more" [11].

Focus on BSM discoveries has tangible implications. Instead of randomly selecting what features a new detector might have, which a revolutionary or SM falsification orientation would necessarily entail, a BSM-centered approach makes deliberate and justified choices in order to make BSM exclusion/confirmation. An example is the SHiP detector, which introduces many specialized features for BSM purposes. For example, the report details some key features of SHiP introduced to maximize sensitivity to particular BSM scenarios:

> In addition to the mainstream spectrometer, SHiP is planned to be equipped with a high precision emulsion spectrometer located immediately upstream of the decay vessel. This subdetector will increase the discovery reach by providing sensitivity to re-interactions of long lived particles produced in the dump, and will collect a first high statistics sample of τ-neutrino interactions to test lepton universality. [11]

Such specific design choices with BSM tests firmly in mind are detailed for all the experimental suggestions given.

Furthermore, it is emphasized in this report that the new experimental ideas are complementary to other experiments. They can make BSM exclusion/confirmation discoveries that no other experiment can. This is graphically emphasized in many figures of the report [11], including Fig. 17 in the case of projected sensitivities to dark scalars, where different experiments have coverage of different regions of parameter space. Without a BSM context there can be no such considerations. One could just as easily claim that every experiment ever contemplated is unique (which they are in some way) and thus has just as much claim to become realized as any other experiment. There can be little rational basis for making decisions about any experiment without an explicitly invoked BSM context.

1.8.4 BSM Theory

A straightforward observation from reading reports that hope to influence the European Strategy Update is that BSM exclusion/confirmation discoveries are still central pursuits in the experimental realm. There is recognition, albeit implicitly at times, that any justification of new experimental project necessarily involves understanding and demonstrating its BSM exclusion/confirmation capabilities. There can be little rational basis for any such decision without BSM.

Despite the centrality of BSM physics, there is very little discussion on the need to invest in a balanced BSM theory program. The BSM theory community is not a laboratory with a director general. It is not a single large collaboration with hundreds or thousands of members with a spokesperson. It is a collection of (near) solitary pursuits that are centrally vital to the progress of high-energy physics. A sure way to diminish high-energy physics despite the presence of sufficient financial resources is to allow BSM theory to stagnate and diminish by reducing its priority. It is easy to do since there are no institutionally constructed leaders to keep its value prominently reminded, as there is in large experimental collaborations. One must keep in mind, however, that the stagnation and subsequent diminishment of BSM theory means the diminishment of high-energy experiment (and vice-versa of course), and with the loss of BSM theory comes the loss of defensible rationales for transformative experiment.

1.9 When Does Discovery End?

Since the apotheosis of science hundreds of years there have been conjectures, worries, "proofs", and declarations that discovery has ended or will end soon [105]. It is interesting to consider what conditions must be met to feel confident that scientific discovery has come to an end, using the language of discovery developed here. In this conceptualization, proof that scientific discovery ends is fulfilled when all three "propositions of discovery cessation" are proven to be true.

Propositions of Discovery Cessation

1. There is no prospect for exclusion/confirmation discovery.
2. There is no prospect for SM falsification.
3. There is no prospect for revolutionary discovery.

Proving propositions 2 and 3 appear to be formally impossible. Regarding proposition 3 in particular, we have argued above that seeking revolutionary discoveries outside the context of BSM exclusion is mystical in nature, and the arrival of revolutionary discoveries is rare and can never be guaranteed. There is no theory for the expected rate of revolutionary discoveries, and thus there can be no assessment that we will fail to have one again in the future.

Proposition 1 has been argued above as the key proposition for guaranteeing discovery which directly counters any cessation claims. Thus, a practical discussion of whether one is at the end of discovery can be had by assessing the status of BSM theories within the theory canon. In that spirit, cessation proposition 1 could be true for any number of reasons, including these three "conditions":

Conditions of Discovery Curtailment

(a) The theory canon becomes empty of any viable or motivated BSM theory, including BSM-motivated loci of SM parameter space.
(b) Or, if there are BSM theories in the theory canon, there nevertheless is no idea for how to turn reasonable resources of time, people, and money into an experiment that can make exclusion/confirmation discoveries within the canon.
(c) Or, if there exist BSM theories in the theory canon and there exist reasonable experimental ideas to make exclusion/confirmation discoveries, there nevertheless exist insufficient resources to pursue them, such as no willingness by governments to financially invest in discovery, or no willingness of scientists to invest sufficient time to pursue the future discoveries.

It is abundantly clear that we are far from reaching the discovery curtailment conditions (a) and (b). As the technical design reports of the European Strategy process confirm, there is a multitude of interesting BSM theories within the theory canon that are empirically adequate and which can be excluded/confirmed by future proposed experiments.[15] The high-energy physics endeavor is also far from reaching curtailment condition (c). However, it should be noted that condition (c) is frequently the highest risk when it comes to pursuing discovery in science, and it is one that needs as much diligence staving off as the (a) and (b) conditions, even though (a) and (b) are more enjoyable to pursue by the scientists.

In short, there is no justifiable claim that the end of discovery is now or nigh, and there is no present sign of asymptoting to any of the discovery curtailment conditions. Nevertheless, the diminishing of effort toward constructing and maintaining a vibrant

---

[15] And it should be emphasized that the standard for interest in BSM theories is not that they are guaranteed to be found at the next future experiment if they exist, but rather that they purport to solve a problem or some other claim to expectation, and that they have a reasonable, but not necessarily guaranteed, prospect for their effects to be discerned.

BSM theory canon would work to activate curtailment condition (a), the diminishing of investment in devising experimental ideas and methods (including detector and accelerator development) would work to activate curtailment condition (b), and the abandonment of proposing and lobbying for future state-of-the-art facilities that guarantee exclusion/confirmation discovery would work to activate curtailment condition (c). Staving off the end of discovery requires effort on many complementary fronts as a community.

1.10 Summary

Below is a summary of the main arguments contained in this essay, some of which are rather obvious but need to be articulated for coherence, while others are the result of more fully developed provocations above.

- No discovery happens without the existence of a theory canon populated by the SM and empirically adequate BSM theories.
- The three main direct discovery activities of high-energy physics are theory canon building (i.e., model building), experiment building (i.e., creative design, construction and execution), and analysis, which connects canon theories to experimental possibilities and experimental results to theory expectations.
- Among the three broad types of discovery—confirmation, exclusion, and revolutionary—only exclusion discovery can be guaranteed.
- A BSM exclusion discovery is a significant experimental feat since by definition it requires exclusion beyond what all prior experiment has been able to accomplish.
- BSM confirmation discoveries can only happen by experiments that had guaranteed BSM exclusion capability at the outset.
- BSM falsification claims for decoupling theories (i.e., those whose parameter spaces allow observable predictions arbitrarily close to those of the SM) often require the application of strong non-empirical constraints to their parameter spaces, such as simplicity and lack of finetuning, and thus are of controversial significance and are inherently polemical from the strict empirical perspective.
- BSM falsifications for theories that are defined with no strong non-empirical constraints on their parameter spaces are secure and non-controversial falsification claims, since the parameter space boundaries are not up for controversial non-empirical re-assessments.
- Approval for any future high-investment research facility should be reserved for those with guaranteed discovery, or in other words, with demonstrated capacity to make BSM exclusion discoveries with respect to the theory canon.
- The main methodological rival to exploration centered on BSM exclusion/confirmation is signalism, which aims through "signal-based analysis" and "model independent searches" to achieve SM falsification or revolutionary discoveries without any reference to BSM theories.

- Signalism's approach to pursue only revolutionary discoveries without reference to BSM theories leads to inscrutable mystical exertions that have no rational claim to utility, while at the same time poses unwelcome risks to BSM confirmation/exclusion discoveries and SM falsification.
- Signalism's closely related other approach, to pursue only SM falsification without reference to any BSM theories, poses risks to every type discovery, including SM falsification, while focus on BSM exclusion/confirmation discovery heightens prospects for SM falsification.
- The SM is just one empirically adequate theory among many, and thus the asymmetric worship of the SM and disdain for BSM theories among many signalism-oriented physicists is an ironical manifestation of a rigid extra-empirical philosophy of theory choice.
- The prospects for truly revolutionary discoveries are unlikely to be adversely affected by focus on BSM exclusion/confirmation compared to any other approach hoping to maximize revolutionary discovery's chances.
- Excitement for some recent SM feature confirmations (e.g., Higgs boson and gravity waves) versus others (e.g., new hadronic resonance) is not to be understood by intrinsic SM-based worth criteria, but rather is due to the BSM exclusion/confirmations that are made possible presently and into the future by the result, which further reveals how central BSM is to excitement, interest and intuited notions of progress.
- Discovery hopes cease when *all* of the following three conditions are shown to be true:
  (1) there is no prospect for BSM exclusion/confirmation discoveries;
  (2) there is no prospect for SM falsification discovery; and,
  (3) there is no prospect for revolutionary discoveries.
  We are presently far from justifying any of these three propositions of discovery cessation.
- Discovery can be curtailed when *any* of the following three conditions is present:
  (1) the BSM theory canon is allowed to languish;
  (2) experimental ideas to confirm/exclude the BSM canon diminish; or,
  (3) governmental/institutional support diminishes.
  All three require continual attention to maintain healthy discovery in high-energy physics.

High-energy physics is rather unique among the sciences, as it is by construction a field of inquiry that pursues a frontier that is by all practical definitions infinite, and thus promises mystery and anticipation for as far as the mind can see. As such, its primary goal is to make discoveries akin to sea-faring explorers of the past—a journey complete with financiers, officers, subalterns, visionaries, and mutineers. As pointed out above, discoveries are only possible through the passage of BSM exclusions, just as Ponce de León's discovery of Florida was made possible only through passage of open seas. High-energy physics has expertly charted the high seas for some time, noting the fascinating islands of recent confirmation discoveries that include the charm quark, the top quark, massive neutrinos, and the Higgs boson. We sail further. Maybe there is a near-by continent up ahead. Let's see.

References

1. Aaboud, M., et al. [ATLAS Collaboration]: Search for new phenomena using the invariant mass distribution of same-flavour opposite-sign dilepton pairs in events with missing transverse momentum in $\sqrt{s} = 13$ TeV pp collisions with the ATLAS detector. Eur. Phys. J. C **78**(8), 625 (2018). https://doi.org/10.1140/epjc/s10052-018-6081-9, arXiv:1805.11381 [hep-ex]
2. Aaboud, M., et al. [ATLAS Collaboration]: Measurement of the top quark mass in the $t\bar{t} \rightarrow$ lepton+jets channel from $\sqrt{s} = 8$ TeV ATLAS data and combination with previous results. arXiv:1810.01772 [hep-ex]
3. Aad, G., et al. [ATLAS Collaboration]: Observation of a new particle in the search for the standard model Higgs boson with the ATLAS detector at the LHC. Phys. Lett. B **716**, 1 (2012). https://doi.org/10.1016/j.physletb.2012.08.020, arXiv:1207.7214 [hep-ex]
4. Aaij, R., et al. [LHCb Collaboration]: Observation of two resonances in the $\Lambda_b^0 \pi^{\pm}$ systems and precise measurement of $\Sigma_b^{\pm}$ and $\Sigma_b^{*\pm}$ properties. Phys. Rev. Lett. **122**(1), 012001 (2019). https://doi.org/10.1103/PhysRevLett.122.012001, arXiv:1809.07752 [hep-ex]
5. Aaij, R., et al. [LHCb collaboration]: Observation of CP violation in charm decays. CERN-EP-2019-042 (13 March 2019). http://cds.cern.ch/record/2668357/files/LHCb-PAPER-2019-006.pdf
6. Amsler, C., et al. (Particle Data Group): Review of particle properties. Phys. Lett. **667**, 1 (2008)
7. Abachi, S., et al. [D0 Collaboration]: Observation of the top quark. Phys. Rev. Lett. **74**, 2632 (1995). https://doi.org/10.1103/PhysRevLett.74.2632 [hep-ex/9503003]
8. Abe, F., et al. [CDF Collaboration]: Observation of top quark production in $\bar{p}p$ collisions. Phys. Rev. Lett. **74**, 2626 (1995). https://doi.org/10.1103/PhysRevLett.74.2626 [hep-ex/9503002]
9. Abe, K., et al. [Belle Collaboration]: Observation of large CP violation in the neutral B meson system. Phys. Rev. Lett. **87**, 091802 (2001). https://doi.org/10.1103/PhysRevLett.87.091802 [hep-ex/0107061]
10. Abramowicz, H., et al.: Higgs physics at the CLIC electron-positron linear collider. Eur. Phys. J. C **77**(7), 475 (2017). https://doi.org/10.1140/epjc/s10052-017-4968-5, arXiv:1608.07538 [hep-ex]
11. Alemany, R., et al.: Summary report of physics beyond colliders at CERN. arXiv:1902.00260 [hep-ex]
12. Anderson, C.D.: The positive electron. Phys. Rev. **43**, 491 (1933). https://doi.org/10.1103/PhysRev.43.491
13. Arbey, A., Cacciapaglia, G., Cai, H., Deandrea, A., Le Corre, S., Sannino, F.: Fundamental composite electroweak dynamics: status at the LHC. Phys. Rev. D **95**(1), 015028 (2017). https://doi.org/10.1103/PhysRevD.95.015028, arXiv:1502.04718 [hep-ph]
14. Arkani-Hamed, N., Dimopoulos, S., Dvali, G.R.: The hierarchy problem and new dimensions at a millimeter. Phys. Lett. B **429**, 263 (1998). https://doi.org/10.1016/S0370-2693(98)00466-3 [hep-ph/9803315]
15. Arkani-Hamed, N., Porrati, M., Randall, L.: Holography and phenomenology. JHEP **0108**, 017 (2001). https://doi.org/10.1088/1126-6708/2001/08/017 [hep-th/0012148]
16. Arnison, G., et al. [UA1 Collaboration]: Experimental observation of isolated large transverse energy electrons with associated missing energy at $\sqrt{s} = 540$ GeV. Phys. Lett. B **122**, 103 (1983) [Phys. Lett. **122B**, 103 (1983)]. https://doi.org/10.1016/0370-2693(83)91177-2

17. Atlas Collaboration: Higgs physics results. https://twiki.cern.ch/twiki/bin/view/AtlasPublic/HiggsPublicResults. Accessed 27 Jan 2019

18. Aubert, J.J., et al. [E598 Collaboration]: Experimental observation of a heavy particle J. Phys. Rev. Lett. **33**, 1404 (1974). https://doi.org/10.1103/PhysRevLett.33.1404

19. Aubert, B., et al. [BaBar Collaboration]: Observation of CP violation in the B^0 meson system. Phys. Rev. Lett. **87**, 091801 (2001). https://doi.org/10.1103/PhysRevLett.87.091801 [hep-ex/0107013]

20. Augustin, J.E., et al. [SLAC-SP-017 Collaboration]: Discovery of a narrow resonance in e^+e^- annihilation. Phys. Rev. Lett. **33**, 1406 (1974) [Adv. Exp. Phys. **5**, 141 (1976)]. https://doi.org/10.1103/PhysRevLett.33.1406

21. Azzi, P., et al. [HL-LHC Collaboration and HE-LHC Working Group]: Standard model physics at the HL-LHC and HE-LHC. arXiv:1902.04070 [hep-ph]

22. Baer, H., et al.: The international linear collider technical design report—Volume 2: Physics. arXiv:1306.6352 [hep-ph]

23. Bagnaschi, E., et al.: Supersymmetric models in light of improved Higgs mass calculations. Eur. Phys. J. C **79**(2), 149 (2019). https://doi.org/10.1140/epjc/s10052-019-6658-y, arXiv:1810.10905 [hep-ph]

24. Baldes, I., Servant, G.: High scale electroweak phase transition: baryogenesis and symmetry non-restoration. JHEP **1810**, 053 (2018). https://doi.org/10.1007/JHEP10(2018)053, arXiv:1807.08770 [hep-ph]

25. Banner, M., et al. [UA2 Collaboration]: Observation of single isolated electrons of high transverse momentum in events with missing transverse energy at the CERN $p\bar{p}$ collider. Phys. Lett. B **122**, 476 (1983) [Phys. Lett. **122B**, 476 (1983)]. https://doi.org/10.1016/0370-2693(83)91605-2

26. Barnett, M., Quinn, H.R., Mühry, H.: The Charm of Strange Quarks: Mysteries and Revolutilns of Particle Physics. Springer, New York (2000)

27. Bettini, A.: Problems and status of neutrino physics. In: EMFCSC International School of Subnuclear Physics. Erice, 14–23 June 2018

28. Bezrukov, F.L., Shaposhnikov, M.: The standard model Higgs boson as the inflaton. Phys. Lett. B **659**, 703 (2008). https://doi.org/10.1016/j.physletb.2007.11.072, arXiv:0710.3755 [hep-th]

29. Blondel, A.: The number of neutrinos and the Z line shape. Adv. Ser. Direct. High Energy Phys. **26**, 145 (2016). https://doi.org/10.1142/9789814733519_0008

30. Bloom, E.D., et al.: High-energy inelastic e p scattering at 6-degrees and 10-degrees. Phys. Rev. Lett. **23**, 930 (1969). https://doi.org/10.1103/PhysRevLett.23.930

31. Bogolyubov, N.N., et al. (eds.): Euler and Modern Science. Mathematical Association of America (2007)

32. Breidenbach, M., et al.: Observed behavior of highly inelastic electron-proton scattering. Phys. Rev. Lett. **23**, 935 (1969). https://doi.org/10.1103/PhysRevLett.23.935

33. For a famous illustration of the focus on failure, see Browne, M.W.: 315 Physicists Report Failure in Search for Supersymmetry. New York Times, 5 January 1993

34. Buchmuller, W., Wyler, D.: Nucl. Phys. B **268**, 621 (1986). https://doi.org/10.1016/0550-3213(86)90262-2

35. Brivio, I., Trott, M.: The standard model as an effective field theory. https://doi.org/10.1016/j.physrep.2018.11.002, arXiv:1706.08945 [hep-ph]

36. Caldwell, R.R., Smith, T.L., Walker, D.G.E.: Using a primordial gravitational wave background to illuminate new physics. arXiv:1812.07577 [astro-ph.CO]

37. Caldwell, A., Dvali, G.: On the gravitational force on anti-matter. arXiv:1903.09096 [hep-ph]

38. Carena, M., Da Rold, L., Pontón, E.: Minimal composite Higgs models at the LHC. JHEP **1406**, 159 (2014). https://doi.org/10.1007/JHEP06(2014)159, arXiv:1402.2987 [hep-ph]

39. Cepeda, M., et al. [Physics of the HL-LHC Working Group]: Higgs physics at the HL-LHC and HE-LHC. arXiv:1902.00134 [hep-ph]

40. CERN Press Office. CERN experiments observe particle consistent with long-sought Higgs boson, 4 July 2012. http://cds.cern.ch/journal/CERNBulletin/2012/28/News%20Articles/1459454. Accessed 4 Feb 2019

41. Chatrchyan, S., et al. [CMS Collaboration]: Observation of a new boson at a mass of 125 GeV with the CMS experiment at the LHC. Phys. Lett. B **716**, 30 (2012). https://doi.org/10.1016/j.physletb.2012.08.021, arXiv:1207.7235 [hep-ex]
42. Chatrchyan, S., et al. [CMS Collaboration]: Evidence for the direct decay of the 125 GeV Higgs boson to fermions. Nat. Phys. **10**, 557 (2014). https://doi.org/10.1038/nphys3005, arXiv:1401.6527 [hep-ex]
43. Cheung, C.: TASI lectures on scattering amplitudes. https://doi.org/10.1142/9789813233348_0008, arXiv:1708.03872 [hep-ph]
44. Cid Vidal, X., et al.: Beyond the standard model physics at the HL-LHC and HE-LHC. arXiv:1812.07831 [hep-ph]
45. Close, F.: Antimatter. Oxford University Press (2009)
46. CMS Collaboration. Higgs physics results. https://twiki.cern.ch/twiki/bin/view/CMSPublic/PhysicsResultsHIG. Accessed 27 Jan 2019
47. Csaki, C., Schmaltz, M., Skiba, W.: Confinement in N=1 SUSY gauge theories and model building tools. Phys. Rev. D **55**, 7840 (1997). https://doi.org/10.1103/PhysRevD.55.7840 [hep-th/9612207]
48. Cui, Y., Lewicki, M., Morrissey, D.E., Wells, J.D.: Cosmic archaeology with gravitational waves from cosmic strings. Phys. Rev. D **97**(12), 123505 (2018). https://doi.org/10.1103/PhysRevD.97.123505, arXiv:1711.03104 [hep-ph]
49. Cui, Y., Lewicki, M., Morrissey, D.E., Wells, J.D.: Probing the pre-BBN universe with gravitational waves from cosmic strings. JHEP **1901**, 081 (2019). https://doi.org/10.1007/JHEP01(2019)081, arXiv:1808.08968 [hep-ph]
50. Curtin, D., et al.: Exotic decays of the 125 GeV Higgs boson. Phys. Rev. D **90**(7), 075004 (2014). https://doi.org/10.1103/PhysRevD.90.075004, arXiv:1312.4992 [hep-ph]
51. Davoudiasl, H., Kitano, R., Li, T., Murayama, H.: The new minimal standard model. Phys. Lett. B **609**, 117 (2005). https://doi.org/10.1016/j.physletb.2005.01.026 [hep-ph/0405097]
52. Dawson, S., Englert, C., Plehn, T.: Higgs physics: it ain't over till it's over. arXiv:1808.01324 [hep-ph]
53. Degrassi, G., Di Vita, S., Elias-Miro, J., Espinosa, J.R., Giudice, G.F., Isidori, G., Strumia, A.: Higgs mass and vacuum stability in the standard model at NNLO. JHEP **1208**, 098 (2012). https://doi.org/10.1007/JHEP08(2012)098, arXiv:1205.6497 [hep-ph]
54. Denegri, D., Sadoulet, B., Spiro, M.: The number of neutrino species. Rev. Mod. Phys. **62**, 1 (1990). https://doi.org/10.1103/RevModPhys.62.1
55. Diehl, E., Kane, G.L., Kolda, C.F., Wells, J.D.: Theory, phenomenology, and prospects for detection of supersymmetric dark matter. Phys. Rev. D **52**, 4223 (1995). https://doi.org/10.1103/PhysRevD.52.4223 [hep-ph/9502399]
56. Dirac, P.A.M.: The quantum theory of the electron. Proc. R. Soc. Lond. A **117**, 610 (1928). https://doi.org/10.1098/rspa.1928.0023
57. Draper, P., Rzehak, H.: A review of Higgs mass calculations in supersymmetric models. Phys. Rep. **619**, 1 (2016). https://doi.org/10.1016/j.physrep.2016.01.001, arXiv:1601.01890 [hep-ph]
58. Eagleton, T.: Literary Theory, 2nd edn. University of Minnesota Press, Minneapolis (1996)
59. Elvang, H., Huang, Y.T.: Scattering amplitudes. arXiv:1308.1697 [hep-th]
60. European Strategy Update. The European Particle Physics Strategy Update 2018–2020. http://europeanstrategyupdate.web.cern.ch/. Accessed 17 Jan 2019
61. Farrington, B.: Science in Antiquity, 2nd edn. Oxford University Press (1969)
62. Feynman, R.: The Character of Physical Law. MIT Press (1965)
63. Foucault, M.: What is an author? (Trans. by J.V. Harari). In: Rabinow, P. (ed.) The Foucault Reader. Pantheon, New York (1984)
64. Fujii, K., et al.: Physics case for the 250 GeV stage of the international linear collider. arXiv:1710.07621 [hep-ex]
65. Fukuda, Y., et al. [Super-Kamiokande Collaboration]: Measurement of the flux and zenith angle distribution of upward through going muons by Super-Kamiokande. Phys. Rev. Lett. **82**, 2644 (1999). https://doi.org/10.1103/PhysRevLett.82.2644 [hep-ex/9812014]

66. Garisto, R.: Focus: neutrinos have mass. APS Phys. (1998). https://physics.aps.org/story/v2/st10. Accessed 17 Dec 2018

67. Gilmer, P.J.: Irène Joliot-Curie, a Nobel laureate in artificial radioactivity. In: Chiu, M.-H., Gilmer, P.J., Treagust, D.F. (eds.) Celebrating the 100th Anniversary of Madame Marie Sklodowska Curie's Nobel Prize in Chemistry. Sense Publishers, Rotterdam (2011)

68. Giudice, G.F.: Naturally speaking: the naturalness criterion and physics at the LHC. arXiv:0801.2562 [hep-ph]

69. Giudice, G.F.: On future high-energy colliders. arXiv:1902.07964 [physics.hist-ph]

70. Grojean, C., Servant, G., Wells, J.D.: First-order electroweak phase transition in the standard model with a low cutoff. Phys. Rev. D **71**, 036001 (2005). https://doi.org/10.1103/PhysRevD.71.036001 [hep-ph/0407019]

71. Harrison, P.F., Perkins, D.H., Scott, W.G.: Tri-bimaximal mixing and the neutrino oscillation data. Phys. Lett. B **530**, 167 (2002). https://doi.org/10.1016/S0370-2693(02)01336-9 [hep-ph/0202074]

72. Hatakeyama, S., et al. [Kamiokande Collaboration]: Measurement of the flux and zenith angle distribution of upward through going muons in Kamiokande II + III. Phys. Rev. Lett. **81**, 2016 (1998). https://doi.org/10.1103/PhysRevLett.81.2016 [hep-ex/9806038]

73. Hoddeson, L., Brown, L., Riordan, M., Dresden, M. (eds.): The Rise of the Standard Model. Cambridge University Press (1997)

74. Jarlskog, C.: Public comment, Lund October 2014

75. Kennefick, D.: Einstein versus the physical review. Phys. Today **58**(9), 43 (2005). https://doi.org/10.1063/1.2117822

76. Khachatryan, V., et al. [CMS Collaboration]: Measurement of the top quark mass using proton-proton data at $\sqrt{(s)}$ = 7 and 8 TeV. Phys. Rev. D **93**(7), 072004 (2016). https://doi.org/10.1103/PhysRevD.93.072004, arXiv:1509.04044 [hep-ex]

77. King, S.F.: Neutrino mass models. Rept. Prog. Phys. **67**, 107 (2004). https://doi.org/10.1088/0034-4885/67/2/R01 [hep-ph/0310204]

78. Lee, E.R., Halyo, V., Lee, I.T., Perl, M.L.: Automated electric charge measurements of fluid microdrops using the Millikan method. Metrologia **41**, S147 (2004). https://doi.org/10.1088/0026-1394/41/5/S05

79. Lipkin, H.J.: Theory, phenomenology, and 'who ordered that?' Phys. Today **54**(1), 68 (2001). https://doi.org/10.1063/1.4796210

80. Maldacena, J.M.: The large N limit of superconformal field theories and supergravity. Int. J. Theor. Phys. **38**, 1113 (1999) [Adv. Theor. Math. Phys. **2**, 231 (1998)]. https://doi.org/10.1023/A:1026654312961, https://doi.org/10.4310/ATMP.1998.v2.n2.a1 [hep-th/9711200]

81. Martin, S.P.: A supersymmetry primer. Adv. Ser. Direct. High Energy Phys. **21**, 1 (2010) [Adv. Ser. Direct. High Energy Phys. **18**, 1 (1998)]. https://doi.org/10.1142/9789812839657_0001, https://doi.org/10.1142/9789814307505_0001 [hep-ph/9709356]

82. Moortgat-Pick, G., et al.: Physics at the e^+e^- linear collider. Eur. Phys. J. C **75**(8), 371 (2015). https://doi.org/10.1140/epjc/s10052-015-3511-9, arXiv:1504.01726 [hep-ph]

83. Murayama, H., Pierce, A.: Not even decoupling can save minimal supersymmetric SU(5). Phys. Rev. D **65**, 055009 (2002). https://doi.org/10.1103/PhysRevD.65.055009 [hep-ph/0108104]

84. Pais, A.: Inward Bound. Oxford University Press (1986)

85. Panico, G., Wulzer, A.: The Composite Nambu-Goldstone Higgs. Lecture Notes in Physics. Springer (2016). See also arXiv:1506.01961

86. Patt, B., Wilczek, F.: Higgs-field portal into hidden sectors. [hep-ph/0605188]

87. Perl, M.: Popular and unpopular ideas in particle physics. Phys. Today **39**, 12 (1986)

88. Perlmutter, S., et al. [Supernova Cosmology Project Collaboration]: Discovery of a supernova explosion at half the age of the Universe and its cosmological implications. Nature **391**, 51 (1998). https://doi.org/10.1038/34124 [astro-ph/9712212]

89. Perlmutter, S., et al. [Supernova Cosmology Project Collaboration]: Measurements of omega and lambda from 42 high redshift supernovae. Astrophys. J. **517**, 565 (1999). https://doi.org/10.1086/307221 [astro-ph/9812133]

90. Poulin, V., Smith, T.L., Grin, D., Karwal, T., Kamionkowski, M.: Cosmological implications of ultralight axionlike fields. Phys. Rev. D **98**(8), 083525 (2018). https://doi.org/10.1103/PhysRevD.98.083525, arXiv:1806.10608 [astro-ph.CO]
91. Ramond, P.: Neutrinos and particle physics models. Presented at History of the Neutrino, Paris, September 2018. arXiv:1902.01741 [physics.hist-ph]
92. Randall, L., Sundrum, R.: A large mass hierarchy from a small extra dimension. Phys. Rev. Lett. **83**, 3370 (1999). https://doi.org/10.1103/PhysRevLett.83.3370 [hep-ph/9905221]
93. Redmond, K., Trezza, A., Erickcek, A.L.: Growth of dark matter perturbations during kination. Phys. Rev. D **98**(6), 063504 (2018). https://doi.org/10.1103/PhysRevD.98.063504, arXiv:1807.01327 [astro-ph.CO]
94. Riess, A.G., et al. [Supernova Search Team]: Observational evidence from supernovae for an accelerating universe and a cosmological constant. Astron. J. **116**, 1009 (1998). https://doi.org/10.1086/300499 [astro-ph/9805201]
95. Roseveare, N.T.: Mercury's Perihelion: From Le Verrier to Einstein. Clarendon Press, Oxford (1982)
96. APS Sakurai Prize for Theoretical Particle Physics. https://www.aps.org/units/dpf/awards/sakurai.cfm. Accessed 16 Feb 2019
97. Schabinger, R.M., Wells, J.D.: A minimal spontaneously broken hidden sector and its impact on Higgs boson physics at the large hadron collider. Phys. Rev. D **72**, 093007 (2005). https://doi.org/10.1103/PhysRevD.72.093007 [hep-ph/0509209]
98. Schael, S., et al. [ALEPH and DELPHI and L3 and OPAL and SLD Collaborations and LEP Electroweak Working Group and SLD Electroweak Group and SLD Heavy Flavour Group]: Precision electroweak measurements on the Z resonance. Phys. Rep. **427**, 257 (2006). https://doi.org/10.1016/j.physrep.2005.12.006 [hep-ex/0509008]
99. Seiberg, N.: Electric-magnetic duality in supersymmetric nonAbelian gauge theories. Nucl. Phys. B **435**, 129 (1995). https://doi.org/10.1016/0550-3213(94)00023-8 [hep-th/9411149]
100. Senjanovic, G.: Proton decay and grand unification. AIP Conf. Proc. **1200**, 131 (2010). https://doi.org/10.1063/1.3327552, arXiv:0912.5375 [hep-ph]
101. Sirunyan, A.M., et al. [CMS Collaboration]: Measurement of the top quark mass in the all-jets final state at $\sqrt{s} = 13$ TeV and combination with the lepton+jets channel. arXiv:1812.10534 [hep-ex]
102. Tanabashi, M., et al. (Particle Data Group): Review of particle properties. Phys. Rev. **D98**, 030001 (2018)
103. Veltman, M.J.G.: The infrared-ultraviolet connection. Acta Phys. Polon. B **12**, 437 (1981)
104. Wells, J.D.: Naturalness, extra-empirical theory assessments, and the implications of skepticism. Found. Phys. (2018). https://doi.org/10.1007/s10701-018-0220-x, arXiv:1806.07289 [physics.hist-ph]
105. Wells, J.D.: Prof. von Jolly's 1878 prediction of the end of theoretical physics. In: Essays & Commentaries I. Ann Arbor, MI (2016). https://deepblue.lib.umich.edu/handle/2027.42/148318
106. Wells, J.D.: Beyond the hypothesis: theory's role in the genesis, opposition, and pursuit of the Higgs boson. Stud. Hist. Philos. Mod. Phys. B **62**, 36 (2018). https://doi.org/10.1016/j.shpsb.2017.05.004
107. Wells, J.D.: Finetuned cancellations and improbable theories. To be published in Found. Phys. arXiv:1809.03374 [physics.hist-ph]
108. Williams, P.: Naturalness, the autonomy of scales, and the 125 GeV Higgs. Stud. Hist. Philos. Sci. B **51**, 82 (2015). https://doi.org/10.1016/j.shpsb.2015.05.003
109. Wolfenstein, L.: Theory, phenomenology, and 'who ordered that?' Phys. Today **54**(1), 13 (2001). https://doi.org/10.1063/1.1349597
110. Wu, C.S., Ambler, E., Hayward, R.W., Hoppes, D.D., Hudson, R.P.: Experimental test of parity conservation in beta decay. Phys. Rev. **105**, 1413 (1957). https://doi.org/10.1103/PhysRev.105.1413

Chapter 2
The Once and Present Standard Model of Elementary Particle Physics

Abstract There are many theories that have resided these last fifty years within the hazy mist we have been calling the Standard Model (SM) of elementary particles. An attempt is made here to construct a coherent description of the SM today, because only precisely articulated theories can be targeted for annihilation, corroboration, and alteration. To this end it is useful to categorize the facts, mysteries and myths that together build a single conception of the SM. For example, it is argued that constructing a myth for how neutrinos obtain mass is useful for progress. We also advocate for interpreting the cosmological constant, dark matter, baryogenesis, and inflation as four "mysteries of the cosmos" that are indeterminate regarding new particles or interactions, despite a multitude of available particle explanations. Some history of the ever-changing SM is also presented to remind us that today's SM is not our parents' SM, nor will it likely be our children's SM.

2.1 Introduction

We have known that neutrinos have mass for over two decades, and we had theoretically and experimentally built support for the case that neutrinos had mass for several decades prior to that. Yet, we continue to say phrases like "neutrinos are massless in the Standard Model."[1] This is certainly not out of ignorance, since it is being said by outstanding scientists who are not confused. An underlying reason for this is because we as a community have never really confronted precisely what we mean when we say "Standard Model." Does it mean what physicists envisioned it to mean in 1974, and thus is a static definition tied to predilections at the beginning of modern particle physics (no neutrino masses, no third generation, shakiness on

---

[1]For example, "In the Standard Model (SM) of elementary particle physics, neutrinos are massless particles" [65]. And, "Some considered [the Higgs boson] as the last brick in the construction of the Standard Model. It is not, since in the Standard Model neutrinos have no mass..." [61]. And, "The standard model of particle physics says neutrinos should be massless, but experiments have shown that they have a small but nonzero mass—the subject of the 2015 Nobel Prize in Physics" [34]. There are many more such quotes throughout the literature and presented in talks.

© The Author(s), under exclusive license to Springer Nature Switzerland AG 2020   51
J. D. Wells, *Discovery Beyond the Standard Model of Elementary Particle Physics*,
SpringerBriefs in Physics, https://doi.org/10.1007/978-3-030-38204-9_2

Higgs boson, rising premature acceptance of grand unification, etc.)? Or, is "Standard Model" a dynamic name that is equivalent to "current standard theory" of particle physics, which continually updates itself over time to incorporate the community's current view of the most favored and agreed-upon description of elementary particle physics (including neutrino masses, etc.)? If it is the old static former definition, then it is uninteresting to use the phrase "Standard Model" ever again, except in nostalgic history books, and if it is the new dynamic latter definition then we should not speak of neutrino masses being beyond the Standard Model since it implies we are unable as a community to incorporate that fact into a theoretical structure. Of course, the lack of an agreed upon module for incorporating neutrino masses is at the origin of this confusion with the word, and should be acknowledged. Nevertheless, it is time we cease using the phrase "Standard Model" for a theory we know to be incorrect.

In this article, we advocate for the more useful definition of "Standard Model" (SM), meaning the "current standard theory" of elementary particle physics. In this sense there have been "Standard Models" well before there were quarks and leptons. In the early 20th century it was thought to be entirely made of electrons and protons, and then neutrons were added, and then an explosion of other discoveries happened (leptons, quarks, etc.) bringing us to the modern age.

One can reasonably argue that the modern age of particle physics started in 1974 with the discovery of charm. This "November revolution" [15, 20] was the final offensive that forced all competent resistance to surrender to the $SU(3)_c \times SU(2)_L \times U(1)_Y$ renormalizable gauge theory with elementary quarks and leptons. Only the Higgs boson thereafter faced significant opposition by competent experts before winning the day in 2012 [63]. The SM in place in 1974, however, is not the SM that is in place now. By this I do not mean only that the gauge couplings have been measured better and the Higgs boson has been experimentally confirmed. In other words, points in continuous parameter space measured continually better but not violating any a priori assumed theory structure do not constitute a dismissal of one SM for another. Nor is the confirmation of an elementary particle that had not yet been seen but was *part of the defined theory* a strong reason to say that the original SM is not the same as today's SM. What is valid to say is when there is a qualitatively new phenomenon that was not anticipated in the original SM formulation, and has been corroborated by experiment and understood theoretically, then the old SM must be replaced by a new SM, and it should be recognized as such.

One is therefore drawn to attaching dates to the SM, just as one attaches dates to substantial revisions of a computer programming language (Fortran 77, Fortran 90, etc.). The SM in 1974, which one could call SM-1974, is certainly not the SM of today, SM-2019. They differ by incorporation of third generation, and the acceptance of neutrino masses, and perhaps other ways in which SM-2019 may be defined more precisely. Further defining features may include assumptions on non-renormalizable operators, grand unification, energy domain of applicability, dark matter, θ_{QCD} value for strong CP, etc. We discuss these issues more below, with a major goal of stating what is required to define a SM and what is our SM today.

There is value in articulating very precisely what is the standard theory of some particular domain, such as the SM of particle physics. Precise articulation increases understanding and precision of claims made; it enables clarity on progress in theory development; it forces one to confront areas of ambiguity when equally minimal/simple sectors of the theory compete to be incorporated within the SM (particularly relevant for neutrino physics and dark matter); and, it provides a very clear target for organizing attempts to kill the theory. Fuzzy theories are harder to falsify, and when a theory is susceptible to eliminating its fuzziness, it should do so. The SM is most certainly capable of tightening up its definition, especially within the sector of neutrino physics, as a useful step toward killing it. That is one of our primary aims below.

2.2 Facts, Mysteries and Myths

When working out a theoretical picture for natural phenomena it is useful to separate out facts, mysteries and myths. I will refer to facts as data and conceptual categories that are not presently under question by anyone in the community. For example, the existences of electrons, muons, and $W^{\pm}$ bosons are facts. The realization that neutrinos have mass is another fact. I am ignoring some philosophical subtleties with this free use of the word "facts" but it is adequate for our working purposes here.

Mysteries are questions that we think are important which we cannot answer at the present but we believe there is more information "out there" that can elucidate them in part or in whole. This would include the plausible discovery of new particles or new interactions that directly answer the questions, or at least elucidate them at a deeper level, or it even could include understanding new concepts that remove the question from further interest. Examples of mysteries today are, why is the weak scale so far below the Planck scale? Does the converging behavior of gauge couplings at the high scale signify unification of the forces? How do quantum mechanics and general relativity co-exist peacefully in a unified theoretical structure? What is the origin of three generations of quarks and leptons? What is the origin of four-dimensional spacetime? What is the origin of the phenomena that supports the existence of dark matter? Why is there more matter than anti-matter? There are many additional such questions. The SM has no strong claim to the answers to those questions, yet ideas exist that appear to be satisfactory theoretically and empirically. In Sect. 2.5 I will focus on the cosmological questions as the key mysteries of the SM today.

Myths are mysteries that have been answered concretely, yet the answers have a reasonable probability of being wrong, incomplete or naive. Nevertheless, the community accepts the myths for several useful reasons: they are easy to understand, they elevate the mystery in the consciousness of the community, and perhaps most importantly they enable interpretation of phenomena in a well articulated manner. Before it was discovered in 2012, the Higgs boson fully qualified as a myth within the SM. Some might even say that it continues to be myth if one speaks of the Higgs boson as purely the SM-conceived Higgs boson without any possible deviations, but that would be stretching the myth concept. Here, we can put Higgs boson in the fact column now, making it part of the firm SM, and move on. However, myths with respect to the SM still exist. For example, below I will advocate a specific articulated scenario as a useful myth of how neutrinos obtain their mass.

Therefore, we can define the SM as possessing facts (its known particles and interactions), mysteries (e.g., the origin of matter asymmetry), and myths (e.g., how neutrinos get their masses). Beyond the SM (BSM) physics primarily concerns itself with articulating the mysteries and myths, developing concrete answers to the mysteries, and identifying phenomena and experiment that can shed light on them.

Articulating a tentative definition of a more complete minimal SM was also contained in other works over the last few years [10, 23]. In our language, those works proposed a much richer myth structure than what will be advocated here. These were very useful and laudable exercises. The SM-2004 theory presented in [23], for example, was economical and certainly valid and intriguing at the time. However, it may not be the most minimal in current eyes, and recent experimental developments have put stress on some of the ideas and perhaps point to a different standard choice to make for the SM's relationship to cosmology in particular. That is partly why I prefer putting the many cosmological conundrums in the mysteries column of the SM rather than proposing specific myths for their resolutions. Such considerations will be explained in more detail below.

2.3   Requirements for a Standard Theory

A well known challenge of science is called the "underdetermination problem", which suggests that there is usually more than one theory that can accommodate a set of experimental results [38]. Our experience within particle physics suggests to us an even stronger claim, which we can call the "infinite underdetermination problem" (IUP), which states that there are an infinite number of theories that can accommodate a finite number of imprecise observables. By "imprecise observable" we mean measurements that lead to a determination of an observable with non-zero error (for cross-sections, branching ratios, etc.).

The evidence for IUP is compelling. There are an infinite number of sets of higher-order operators (i.e., arising from an infinite number of unique high energy theories) that decouple from low energy phenomena, yet give tiny shifts in values below error bars of current measurements. Among this infinite number of theories, there is usually a theory class that rises to the top among community researchers because it features a number of desired virtues, such as simplicity, calculability, consilience, unification and of course consistency with experiment. The theory that rises to the top is the "standard theory" for the domain in question.

For particle physics, the standard theory has been called the Standard Model (SM). As we discussed above, we will attach the name SM to the current standard theory of elementary particles and their interactions, as opposed to viewing it as a name for the standard theory that was in place in the early 1970s. Yet, we must ask, does the SM satisfy all the requirements of a standard theory, and if not, what more must we specify?

The standard theory must be (1) a precisely articulated physics theory that is (2) recognized to be the leading theory among the community of scholars. Most would agree that the SM satisfies (2), which is ironic because one usually wants to know what they are voting for, which is impossible to do because the SM comes up short on (1).

To specify what the SM is, one must first decide what we are asking of it as a theory. A physics theory (T) is a set of rules that maps input parameters $\{\xi_i\}$ to experimental observables $\{\mathcal{O}_k\}$ over an agreed-upon target domain $\Delta_{\mathcal{O}}$ of possible observables. Symbolically we can say

$$T : \{\xi_i\} \longrightarrow \{\mathcal{O}_k\} \text{ where all } \mathcal{O}_k \in \Delta_{\mathcal{O}}.$$

The definition of an observable can be subtle, since we are used to observables being defined within the theory framework. For example, the cross-section $e^+e^- \to t\bar{t}$ requires us to have a conception of what an electron is and what a top quark is, which is only provided by the theory itself. Thus, there is some inevitable circularity in the definition of a theory, but that circularity is put under the stringent test of experimental and observational self-consistency. There are ways to reduce that circularity, but such efforts would not be of much practical value in our discussion here. We assume here that there is an intuitive, yet ultimately precise, understanding of what observables are, and what a domain for observables can be.

A standard theory of elementary particles should by definition be a theory of all the putative elementary particles (indivisible) and their interactions. On the surface that is the easy part. One could say that the full set of elementary particles are the fermions (quarks, leptons, and neutrinos), the force carrier vector bosons (photon, $W^\pm$, Z^0, and graviton) and the Higgs boson. Now, regarding the particle content of the SM, if we are content with such imprecision (we should not be), we end up lazily not recognizing many things that have happened over the years since the original SM's birth in 1974. More will be said below about tightening the theory discussion of particle content.

Regarding the target domain of observables ($\Delta_{\mathcal{O}}$), that is also subtle, which is related to the subtlety of defining observables discussed above. If we believe that we have a theory of all elementary particles and all the forces that apply, then in principle we have a "final theory" since everything takes place ultimately at the elementary particle level. Thus, we should in principle be able to not only predict the lifetime of the top quark, but we should also be able to predict the next earthquake. Yet, earthquakes are not within the observables target domain $\Delta_{\mathcal{O}}$ for a standard theory of particle physics, for reasons that are well-known and do not need to be reviewed here. Such examples are not the origin of the target domain subtlety, but rather what energy ranges do we assume the theory to be valid, and in general, what conditions must hold within the target domain of observables.

To this end, we can define the SM more precisely to be a theory ($SU(3) \times SU(2) \times U(1)_Y$ gauge theory) of elementary particle content (quarks, leptons, neutrinos and force carriers) with a parameter space of inputs (gauge couplings, Yukawa couplings, etc.) that enable unambiguous computations of decay lifetime observables ($\Gamma_i \subset \Delta_{\mathcal{O}}$)

and interaction cross-section observables ($\sigma_{ab...} \subset \Delta_{\mathcal{O}}$) in non-extreme gravitational environments (further restriction on $\Delta_{\mathcal{O}}$), where computed results are all consistent within experimental uncertainty for at least one point in the parameter space. By non-extreme gravitational environments we mean when momentum transfers in parton-level collisions are small compared to the Planck mass of $\sim 10^{18}$ GeV, or stated more generally, when the uncertainty of our understanding of strong gravity does not obviously get in the way of computability.[2]

With this more comprehensive definition of what we require out of a standard theory of particle physics, we investigate how the SM has changed since its inception and point out how our current usage of the word "Standard Model" is foggy, and we present a suggestion for making it more precise in a way that satisfies the demands of a standard theory. We require that our more precisely defined SM be within the foggy domain of what is currently meant by the SM, and that it have a strong prospect for being falsified by near-future investigations. And lastly, it should be noted, as with all standard theories in any domain, acceptance of it as the standard theory in no way commits a physicist to believe that it is the "right" or "correct" theory that will remain valid forever in the face of all future theoretical and experimental stresses put to it over time. It merely is designated as the standard theory among the infinite number of currently viable theories that has maximal theory virtues valued by the community at this moment in time. In other words, defining the SM more precisely in no way should be interpreted as heightened arrogance that we know exactly what nature has chosen. Rather, it is a tool through which we track our understanding through time and create firm targets to attack theoretically and experimentally.

2.4   Historical Progression of the Standard Model

In the previous section it was stated that a precise statement on the particle content is required for the SM. It should be kept in mind that there is a difference between when a particle entered the standard expectations of the community, and thus was incorporated within the SM, and when the particle was actually confirmed by experiment. In other words, when tracking the evolution of the SM (i.e., the accepted standard theory of expectations) with respect to particle content, one should focus more on when particles were expected and not on when they were discovered.

For example, the Higgs boson is viewed as one of the most revolutionary discoveries in particle physics in the last fifty years, and rightly so. However, it has been a part of the SM since the beginning. What made it so momentous is that it is a qualitatively new type of elementary particle—a spin-zero boson—that had never been discovered before. As such, it was highly controversial and many competent

---

[2]It is generally held that any particle physics theory based on standard quantum field theory will break down in extreme gravity environments, which is one of the motivations for pursuing deeper string theory descriptions for that domain. It is also why we restrict our discussion to energies well below the Planck scale.

experts had strong suspicions that it *could not exist* even up to the time of its discovery [63]. Nevertheless, it was already part of the SM as an accepted myth—the expectations within the standard theory. Thus, it is not the Higgs boson that has been the source of change over the years to upend one SM in favor of a new SM, despite the extraordinary impact its confirmation discovery has made on science.

Fermion Generations

What has changed is our conception of fermion generations and our conception of the neutrino sector. Let us look at the fermion generation question first. When the SM emerged out of the Glashow-Salam-Weinberg model of electroweak interactions, complete with spontaneous symmetry breaking from the Higgs boson, and combined with the new understanding of quarks and QCD, there was not initially an understanding that there were more than two generations of fermions.

The understanding of the need for three generations came from theory and experiment. In theory work, it was suggested correctly by Kobayashi and Maskawa [37] that a third generation of fermions is needed in order to accommodate CP violation in the kaon system if its origin is through weak interactions. Two generations would not enable complex phase (i.e., CP violation) in what we today call the Cabibbo-Kobayashi-Maskawa (CKM) matrix, but three or more generations do. Later, the third generation was established by experiment over a range of lepton and hadron collider experiments that discovered the tau lepton, bottom quark, tao neutrino, and top quark over a twenty year period, ending in the top's discovery in 1995.

The Original Standard Model (SM-1974)

In the original SM, which we can perhaps call SM-1974, there were only two generations of fermions, with only the charm quark missing. In J. D. Bjorken's 1984 recollections of those early times he described a great uncertainty and fogginess about what really was the underlying physics. He recalled that John Ellis had given a summary talk at an international particle physics conference in London in mid-1974 summarizing all the theory interpretations of the data. As Bjorken remembered it,

> Ellis' catalog well reflected the state of theoretical confusion and general disarray in trying to interpret e^+e^- data. But in the midst of all of this was a talk by John Iliopoulos .... With passionate zealotry, he laid out with great accuracy what we call the standard model. Everything was there: proton decay, charm, the GIM mechanism of course, QCD, the $SU(2) \times U(1)$ electroweak theory, $SU(5)$ grand unification, Higgs, etc. It was all presented with absolute conviction and sounded at the time just a little mad, at least to me (I am a conservative). So at London the pressure to search for charm was there. But even so this was immersed in a rather large degree of confusion. [15]

Bjorken proceeds to describe the confusion regarding experimental attempts to confirm the existence of the second generation charm quark, but then the revolution happened. In Bjorken's words:

> That brings us up then to November 1974. The stage was really set. The balance had changed, and the November revolution just set everything into motion toward the standard model that we have now. Most high energy physicists will probably remember where they were when they first heard about the psi [J/ψ charm meson]. It is like the moon landing, Pearl Harbor

or the Kennedy assasination. I was home and it was dinner hour. Burt Richter called me up and told me the basic parameters over the phone. He said three GeV. I said three GeV per beam, right? He said no, three GeV in the center of mass. I couldn't believe such a crazy thing was so low in mass, was so narrow, and had such a high peak cross-section. It was sensational. [15]

Indeed it was sensational for all the physicists as described by Bjorken here and by others elsewhere (see also [20]).

Unification and the Standard Model

It is interesting to be reminded by the first Bjorken quote above that in the particle physics community's eyes, from the mid 1970s to early 1980s, $SU(5)$ unification, with its generic prediction of proton decay, was extraordinarily compelling. One might even be tempted to put $SU(5)$ unification within the SM-1974 definition, which was then dropped later from the ambient SM mindset after initial experiments looking for proton decay in the early 1980s did not find it. That is a refinement that would be interesting to describe further, but for our purposes here we wish to merely describe the birth of the standard theory, give a feel for how often and substantively the views of the standard theory have changed over the years, and define a current SM that reflects community sentiment.

Neutrino Masses: Disbelief

Coming back to the original SM theory of 1974, we note that it only had two generations of fermions, except for the latent hint from Kobayashi-Maskawa's 1973 work [37] that CP violation can be achieved in weak interactions by a third generation of quarks. In addition to uncertainty about the number of generations assumed in the SM, there was little questioning that the neutrino masses were most likely zero, and thus zero mass neutrinos were a cornerstone of the SM definition. The SM did not have, therefore, right-handed neutrinos ν_R in the spectrum, nor did it recognize or allow for the possibility of the dimension-five Weinberg operator of left-handed neutrinos and Higgs boson $(LH)^2/\Lambda$, where Λ is required to be substantially higher than the weak scale due to the extreme lightness of the neutrino masses compared to other known masses. It is not as though they could not conceive of neutrinos having the possibility of being massive. They did (see, e.g., [14, 39, 42]). It was merely the case of having no compelling evidence for neutrino masses, yet having evidence that if they did exist they had to be many orders of magnitude below the mass of all other known elementary particles. This suggested that it was better to be zero than bizarrely and unexplainably low.

It was not until 1979/1980 that the possibility of neutrinos having mass started gaining widespread community traction. This was the time when the neutrino see-saw became widely known and appreciated within the community, which gave a good reason why neutrino masses could be naturally very tiny compared to the other leptons. Experimental searches were also underway, and first signs of neutrino oscillations, which implied neutrino masses, became evident in Ray Davis's pioneering Homestake experiment in the late 1960s and early 1970s which lead to the "solar neutrino problem" [21].

Nevertheless, the "solar neutrino problem" was viewed as inconclusive since it did not detect all the other neutrinos into which it could have oscillated, and there was question as to how well we understood the sun's complex internal processes, etc. [50]. As Ray Davis reported, "My opinion in the early years was that something was wrong with the standard solar model; many physicists thought there was something wrong with my experiment" [22]. For example, Trimble and Reines's 1973 review on the solar neutrino problem states: "The critical problem is to determine whether the discrepancy is due to faulty astronomy, faulty physics, or faulty chemistry" [60]. Nevertheless, the theory and experimental progress that proceeded led the community from the early 1970s to the early 1990s to adiabatically come around to the expectation, not just the possibility, that neutrinos had mass.

To demonstrate the widely held belief even in the 1980s that neutrinos were massless we can refer to Cheng and Li's *Gauge theory of elementary particle physics* published in 1984 [19], which was one of the most widely read advanced particle physics textbooks. It had this to say about neutrino masses:

> We have seen that the standard theory [now with 3 generations] gives a natural explanation for the presence of the Cabibbo angle and CP phases in quark charged currents. Similarly the same theory helps us to understand the absence of such features in the lepton sector; the masslessness of neutrinos implies that these mixings are physically unobservable. [19]

and

> We have already mentioned in §11.3 that the reason why there are no Cabibbo-like mixing angles in the lepton sector of the standard electroweak theory is neutrino mass degeneracy (i.e. all νs have the same mass — zero). This degeneracy means that there is no need to diagonalize the neutrino mass matrix (in fact no mass mass matrix to begin with). [19]

It is fair to say that there was widespread skepticism and even disbelief of neutrino masses even into the 1980s.

Neutrino Masses: Rising Belief

Throughout the late 1980s and early 1990s the mainstream perspective on neutrino masses shifted. For example, the Particle Data Group bi-yearly updates transitioned from statements like, "If one considers the possibility of nonzero masses for neutrinos..." in 1992 [47] to something much more definitive in 1994 about the community's expectation that neutrinos have mass:

> The theoretical perspective concerning neutrino masses has changed considerably over the past 20 years. Before that time, a standard view was that there was no theoretical reason for neutrinos to have masses.... Indeed, even in the literature of the 1970's, one will often find statements asserting that in the standard $SU(2) \times U(1)$ electroweak theory ... the known ... neutrinos are massless. In contrast, in the modern theoretical view ... small but nonzero neutrino masses are expected on general grounds. [48]

Certainly by the early 1990s the community was firmly behind the proposition that neutrinos had mass and it was part of the standard theory. The pressure that took the community from no masses to belief in neutrino masses was due to many factors, including "neutrino anomalies" among several experiments and the rising new theory perspectives of string theories, grand unified theories, and supersymmetry. These theory perspectives liked neutrino masses and were gaining indirect experimental support from LEP precision measurements of the gauge couplings that pointed toward supersymmetric unification, further supporting string theory ideas and by consequence the other intuitions that came with it, including neutrino masses (especially through E_6 unification and its subgroup $SO(10)$). Despite the standard theory never being supersymmetric, nor being described directly as a string theory, their influence extended to the low-energy standard model theory expectations, and thus the expectations of neutrino masses were solidified both experimentally and theoretically, and anomalies began to be interpreted as evidence for mass. For example, the arguments delivered by Robert Shrock in the PDG in 1994 [48] for why non-zero neutrino masses are expected are primarily from a string theory perspective, which reflects the sentiments of the writer who, in his community responsibility as PDG contributor, is presumably summarizing widely held viewpoints. In any event, it is rather safe to say that SM-1990s was a theory with neutrino masses.

Neutrino masses were experimentally beyond reproach finally by 1998. That is when Super-Kamiokande firmly established a self-consistent and comprehensive picture of neutrino oscillations [28]. Ever since there has been no question about its required presence within the SM. The mass differences and mixing angles have been measured with increasing precision over the years since. A new push to measure the neutrino sector even more precisely is underway, including the many current and future flagship programs at Fermilab [61].

Neutrino Masses: Theories

Despite all of this attention on neutrino physics, there is no clear agreement of what the neutrino sector is within the SM. There are many possibilities. One possibility is to add a right-handed neutrino and then add the Yukawa operator

$$\mathscr{L}_1 \subset y_\nu \bar{L} H \nu_R \quad \text{(Dirac neutrinos)}.$$

These Dirac neutrinos are then given mass according to $m_\nu = y_\nu \langle H \rangle$. Or, it could be defined without adding any new particles, and the masses are generated by the dimension-five Weinberg operator

$$\mathscr{L}_2 \subset \frac{1}{\Lambda} (\bar{L} H)(\bar{L} H) \quad \text{(Majorana neutrinos)}$$

where Λ is a scale much higher than the electroweak scale in order to give tiny Majorana neutrino masses, which are given by $m_\nu = \langle H \rangle^2 / \Lambda$. Or, yet further, one could supplement $\mathscr{L}_1$ with a right-handed neutrino Majorana mass to give

$$\mathscr{L}_3 \subset y_\nu \bar{L} H \nu_R + M \nu_R \nu_R \quad \text{(seesaw)}$$

which, if $M \gg y_\nu \langle H \rangle$, yields light neutrino masses of mostly left-handed composition with mass $m_\nu \sim (y_\nu \langle H \rangle)^2 / \Lambda$. This is the famous seesaw mechanism for generating tiny neutrino masses even if $y_\nu \langle H \rangle$ is on order of other leptons and quarks in the theory [30, 43, 44, 54, 66].

Thus, we have three theories to consider for neutrino masses, all of which co-exist in a fuzzy superposition of what the community refers to when they say SM. In order to define SM-2019 precisely we need to choose which of the many theories is to be designated the standard theory. The two simplest are $\mathscr{L}_1$ and $\mathscr{L}_2$, so we eliminate $\mathscr{L}_3$ from the running. $\mathscr{L}_1$ has the advantage of introducing only three ν_R's along with a corresponding Yukawa coupling matrix y_ν whose entries are additional input parameters to the SM. $\mathscr{L}_2$ has less fields (i.e., no ν_R's) and the same number of input parameters associated with the coefficient matrix to the Weinberg operator. However, it is not a viable theory across the full energy range of interaction strengths that are not gravitationally strongly coupled. For example, scatterings of neutrinos with energies well above Λ yet well below M_{Pl} ($\Lambda \ll E_\nu \ll M_{\mathrm{Pl}}$) are generically expected to be altered, whereas the interactions added in $\mathscr{L}_1$ have no immediate worries for where they will break down and so are well-behaved and calculable all the way up to near M_{Pl}.

Neutrino Masses Within the SM

For the reasons stated above, an excellent candidate for SM-2019—the standard theory of elementary particle physics up to M_{Pl} that includes neutrino masses—is one that incorporates neutrino masses by introducing three ν_R's and Yukawa couplings of it to the left-handed doublets L and Higgs boson, as described by $\mathscr{L}_1$ above.

We can go one step further by considering an important question within neutrino physics regarding whether the neutrino mass eigenstates are organized in a normal hierarchy (NH) or inverted hierarchy (IH). The normal hierarchy suggests that the neutrinos that are most matched with the leptons (i.e., flavor eigenstate overlaps) have the same mass hierarchy as the leptons. In other words, is the heaviest neutrino most overlapping with ν_τ, and is the next massive neutrino most overlapping with ν_μ, and is the lightest neutrino most overlapping with ν_e? Such an expected "normal hierarchy" is consistent with all neutrino data. But another possibility is consistent with the data, which has the mass hierarchy inverted in a topologically disconnected region of the parameter space. Knowing which island of parameter space is the correct island, NH or IH, has profound implications for flavor model building [8] and for prospects to find BSM signals in the neutrino sector (see, e.g., [24, 27, 40]). Thus, it is justifiable to specify one of the islands for SM-2019, while the other is relegated to a BSM possibility. We choose NH. The justification for NH in SM-2019 springs from two additional reasons beyond what we have already discussed. First, it follows the standard hierarchy that we have learned from quarks. Second, although somewhat controversial, the data may already be giving a slight preference for NH according to recent analyses [16, 29, 33, 41, 56, 58], and thus may have more empirical claim to be the choice of SM-2019.

Of course, a precise description of the neutrino sector cannot be compelling at this point, and therefore it must be introduced not as a fact but as a useful myth, whose implications can be compared with experiment and discovery progress can be tracked, as we will discuss next.

Challenging the SM Theory of Neutrino Masses

One of the advantages of a unambiguous designation for neutrino physics, such as that provided by Eq. 2.1 and NH, is that theorists and experimentalists can ask how to falsify and test the theory. And as discussed previous [64] the best way to make progress is to make motivated physics theories for physics beyond the Standard Model that give experimental predictions that are not within the realm of possibility for the SM. BSM theories with respect to SM-2019 predict new possible phenomena for supernova neutrinos, neutrino oscillation experiments, and neutrino-less double beta decay ($0\nu\beta\beta$) experiments. The hard pursuit of these phenomena is the best way to crack SM-2019 on the path to a qualitatively new SM. SM-2019 retains lepton number conservation at the perturbative level and thus new signals of FCNC in the lepton sector, such as $\mu \rightarrow e\gamma$ would signify a breakdown. Discoveries of any new particles or interactions in general would falsify SM-2019. Thus, after fulfilling the pre-requisite of actually defining what the SM means, there is a large class of experiments that could falsify SM-2019 unambiguously in the near future within and outside the world of neutrino experiments.

In addition to experimental pressure that can be placed on SM-2019 there is much theory pressure to apply. For example, how viable or "likely" is it that dimensionless Yukawa couplings associated with neutrinos can be so tiny, $y_\nu < 10^{-12}$? Do UV complete theories, such as string theories, allow for such tiny couplings without other accompanying low-scale phenomena predicted? Within SM-2019 there is a conserved lepton number global symmetry which could be broken by adding Majorana mass terms at the renormalizable level for ν_R. Since SM-2019 does not allow these new terms, how stable is that assumption to our attempts to incorporate particle physics with gravity, where it has been conjectured that global symmetries cannot be invoked to prevent otherwise expected terms [12, 32]? What BSM ideas would make the neutrino mass hierarchies and mixing angles more theoretically appealing, and what experimental or observational consequences do these interesting BSM ideas have? All of these questions are on the table, and all may have resolutions to be found in the coming years.

2.5 Four Mysteries of the Cosmos

Other areas that could conceivably be more precisely well defined in addition to neutrino physics in order to complete our careful designation of the standard theory of particle physics could include new states and interactions that explain the cosmological constant, dark matter, the mechanism that accounts for the preponderance of matter over antimatter, and the mechanism that carried out inflation.

There is no good explanation yet of the cosmological constant. On the other hand, the other three "mysteries of the cosmos" (dark matter, inflation, baryogenesis) have something in common: there are a vast number of adequate, mutually exclusive and even qualitatively different ideas to explain each. Unlike neutrino physics, which has less than a handful of well-disciplined simple theoretical structures to explain their masses and mixings, those three mysteries of the cosmos have almost no practical bound in the number of "good ideas" to account for them. A variant on a common aphorism applicable to circumstances like this might say that when you have dozens of mutually exclusive ideas for why something should be true, it means you have no good idea. One should face this fact. There simply cannot be any credible standard theory choice for any of these four mysteries of the cosmos. There is no Secretariat in these races. Every horse is a million to one.

What to do with these *mysteria scientiae*? We do as we do with any deep mystery and humbly say "we do not know." All ideas are on the table, including all ideas we have not thought of yet. We merely say that with respect to these four mysteries of the cosmos we lay prostrate, waiting for and working toward the day of revelation. In the meantime, in the cradle of these mysteries we humbly cannot elevate any as likely particle explanations within the domain that we have defined for the SM.

In the case of dark matter, which is perhaps the most concrete cosmological mystery to solve, the resolution might be primordial black holes (PBHs) [17], which do not require in and of themselves to extend the SM. However, new particles that give rise to special inflationary potentials that produce the right mass spectrum of PBHs might be required,[3] and the two mysteries of inflation and dark matter may become intertwined [11, 45]. Perhaps even baryogenesis arises from reheating dynamics after inflation, and then three mysteries are intertwined. Or perhaps inflation occurs without the need for new particles, such as Higgs-inflation [13, 25, 52, 57]. We do not know yet. The point is that they are all mysteries at this stage in the sense that not one idea is compelling over other ideas.

What are the practical applications of attaching the "four mysteries of the cosmos" to the SM definition? Is this attitude an abdication of scientific explanation? No, it is not abdication of scientific pursuit. Its practical application is to put into the BSM column *any articulated* concrete idea that explains any one of the four mysteries. No concrete explanation for the mysteries can be part of the SM. The SM accepts the mystery. The SM is the mystery. That is yet another reason why a scientist is not content with the SM, and BSM theories must be pursued.

One might worry that accepting the four mysteries of the cosmos as an integral part of the SM definition means that the SM cannot be falsified from any cosmological data now or into the deep far future. That is correct, but it does not mean a new and better SM will not be found. No result of CMB measurements, dark matter distributions, etc., is anticipated that could ever show that the current SM under this definition is wrong. However, one hopes the day will come where enough data is accrued and enough theoretical insight is achieved to produce an articulated theory, which is not the SM (i.e., being that it would have less "mysteries"), that explains the data efficiently and compellingly. That is the day a better and more refined SM

---

[3]See, for example, Sect. IV of [46] for a review of possibilities.

would be born, even though the old SM would remain compatible with all the data since by definition it took no concrete stand.[4]

If such a new compelling and concrete SM can be defined with respect to cosmological evolution, it will likely come complete with new particles and new interactions. Previously, Davoudiasl et al. [23] bravely made concrete choices (canonized "myths" in our language) to explain at least three of the mysteries. There were many who agreed that these choices were among a small set of leading choices of the day, and perhaps it was a legitimate definition of a full SM in 2004 (SM-2004). However, the DM explanation is increasingly strained by LHC and by WIMP DM searches, and its status is very much reduced in many researcher's minds. Furthermore, the simple $m^2\phi^2$ inflationary potential is more or less ruled out now by CMB data.[5] The model would not be put forward today as the SM choice.

At a previous time the community's dominant preference for DM was weakly interacting massive particle (WIMP) near the weak scale. The early days of supersymmetry, especially, gave ascendancy to this DM candidate since it fell out of the supersymmetric spectrum "for free" [36]. However, intrusive searches for WIMP DM have been coming up empty for several decades now [4, 6, 9]. It is certainly not ruled out, but the pressure on the idea has intensified. So what was once probably thought to be the standard theory explanation for DM now has strong additional competitors [55]. A major competitor is the axion, which not only can serve as DM [3, 51], but was invented originally to solve the different problem of suppressing perceived strong sector sources of CP violations [35, 49]. Much work still must go into finding the axion or closing its full window of parameter space. One might be tempted to consider as part of SM-2019 a simple axion model of DM, like [10] has done. But the concern is that it too will be viewed in time as merely the next popular idea for all of DM in a long line of others that were not terribly compelling in absolute terms.

The SM today, it is my claim, should accept the four mysteries of the cosmos, and strive for the day, through experimental and theory work, when the SM no longer looks attractive in the face of a more concrete and compelling BSM theory, which then becomes the new SM with less mystery.

2.6   Summary and Conclusion

In summary, we have argued that the simplest and most conservative (i.e., the least "new" experimental consequences demanded from it) definition of the Standard Model that is adequate for today (SM-2019) is a theory that simultaneously holds

---

[4]Analogously, one recalls that the SM of much of the western world in 325 A.D., as expressed by the Council of Nicaea, held that "God [is] maker of all things both seen and unseen" [59]. SM-325 remains compatible with all the data, but as a theory it has been continually augmented over the years by articulated, computable, and proximate explanatory theories for the mysteries of natural phenomena.

[5]See, for example, Fig. 8 of [7].

the following facts (quarks, leptons and their interactions), mysteries (cosmological mysteries) and myths (neutrino sector):

- Gauge symmetries are $SU(3)_c \times SU(2)_L \times U(1)_Y$, with no additional discrete mod-ing (such as Z_2, Z_3, or Z_6), with their accompanying gauge bosons.
- $SU(2)_L \times U(1)_Y \to U(1)_{em}$ is accomplished by a single Higgs doublet, which gives mass to the $W^\pm$ and Z^0 bosons, and manifests a single propagating elementary scalar, the Higgs boson (h^0).
- The elementary particle content has the gauge bosons (photon, gluons and $W^\pm/Z^0$ weak bosons), three generations of fermions (quarks and leptons), and the Higgs boson. Both left- and right-handed neutrinos are present in the spectrum.
- $B - L$ is a conserved global quantum number, which forbids ν_R Majorana masses.
- $\theta_{QCD} = 0$.
- In the limit of zero gravity, there are only renormalizable interactions among the elementary particles listed above, and all of those interactions must be consistent with the above symmetries and with Poincaré space-time symmetry.
- The neutrino masses are entirely Dirac masses (implication of above conditions), and their masses obey the normal hierarchy (NH) solution.
- The "four mysteries of the cosmos" (cosmological constant, dark matter, inflation, baryogenesis) are accepted as mysteries of the SM without concrete demands for new elementary particles or interactions.

Few would bet their lives on the validity of every line of the SM definition given above, nor would they *on any other* precise formulation that could have been offered. That is what makes particle physics perpetually susceptible to revolution. There are presently many BSM theories that challenge the primacy and stability of the SM formulation above, and there are many current experiments and proposed experiments looking for corresponding new states and new interactions. Nevertheless, the above definition of the SM is presently self-consistent, satisfies all known data, and presents at least as economical structure to explain data and make predictions as any other postulated theory. It can be used to track our progress.

Let us not forget that particle physics has changed rather decisively since the early 1970s, but we have retained the name "Standard Model" throughout it all. This confusion on what exactly is the SM has led some people less versed in the history of particle physics to think that nothing has changed because there is no new name. And, it has led to yet another group of people on the opposite end declaring that we are in a permanent state of beyond the SM because neutrinos have mass. Not recognizing substantial, albeit slow, progress in our evolution of what constitutes the standard theory of elementary particle physics has even had the implicit effect of hypnotizing some into thinking that no noteworthy progress will ever come until spacetime is totally upended and revolutionized through fermionic extra dimensions (supersymmetry), bosonic extra dimensions (Randall-Sundrum, etc.), or manifestations of string excitations. This desensitization to more modest scientific progress, which is relentless yet often not totally surprising when the day of confirmation finally arrives, is connected with diffusing the SM name into a fog across a continent of technically different theories.

   This article has discussed what it means to have a standard theory of elementary particle physics, and it has attempted to motivate the value of being precise about what our standard theory really is at every given moment and perhaps even having labels that change when the standard theory changes. In that sense, the SM of 1974 is different than the SM of 1988, which is different than the SM of other years, and so on, until we reach the SM of today: SM-2019. I believe it is a mistake to think the SM can be just a private choice and is not worth articulating more precisely in a single coherent position. Each of us has heard many private choices that most would not find compelling. Even among the experts there are many who would have liked some time ago for minimal $SU(5)$ GUTs and proton decay operators to be part of the standard theory [15]. Others joked in the mid-1990s that the SM was really the minimal supersymmetric standard model (MSSM). I have tried to give here the most minimal and conservative definition to SM-2019, including identifying its mysteries and proposing a useful myth for the neutrino sector, based not on what I think but on what I think the community's sentiments could plausibly agree to.

   Our standard theory should be articulated often in order to set unambiguous targets for future work and, just as importantly, to track over time our changes of outlook, improvements in understanding, and gains in knowledge. And that is how even "modest" progress can be recognized for what it is: progress.

References

1. Abazajian, K.N., et al. [CMB-S4 Collaboration]: CMB-S4 Science Book, 1st ed. arXiv:1610.02743 [astro-ph.CO]
2. Abi, B., et al. [DUNE Collaboration]: The DUNE Far Detector Interim Design Report Volume 1: Physics, Technology and Strategies. arXiv:1807.10334 [physics.ins-det]
3. Abbott, L.F., Sikivie, P.: A cosmological bound on the invisible axion. Phys. Lett. B **120**, 133 (1983) [Phys. Lett. **120B**, 133 (1983)]. https://doi.org/10.1016/0370-2693(83)90638-X
4. Ackermann, M., et al. [Fermi-LAT Collaboration]: Searching for dark matter annihilation from milky way dwarf spheroidal galaxies with six years of fermi large area telescope data. Phys. Rev. Lett. **115**(23), 231301 (2015). https://doi.org/10.1103/PhysRevLett.115.231301, arXiv:1503.02641 [astro-ph.HE]
5. Agostini, M., Benato, G., Detwiler, J.: Discovery probability of next-generation neutrinoless double-beta decay experiments. Phys. Rev. D **96**(5), 053001 (2017). https://doi.org/10.1103/PhysRevD.96.053001, arXiv:1705.02996 [hep-ex]
6. Akerib, D.S., et al. [LUX Collaboration]: Results from a search for dark matter in the complete LUX exposure. Phys. Rev. Lett. **118**(2), 021303 (2017). https://doi.org/10.1103/PhysRevLett.118.021303, arXiv:1608.07648 [astro-ph.CO]
7. Akrami, Y., et al. [Planck Collaboration]: Planck 2018 results. X. Constraints on inflation. arXiv:1807.06211 [astro-ph.CO]
8. Altarelli, G., Feruglio, F.: Discrete flavor symmetries and models of neutrino mixing. Rev. Mod. Phys. **82**, 2701 (2010). https://doi.org/10.1103/RevModPhys.82.2701, arXiv:1002.0211 [hep-ph]
9. Aprile, E., et al. [XENON Collaboration]: First dark matter search results from the XENON1T experiment. Phys. Rev. Lett. **119**(18), 181301 (2017). https://doi.org/10.1103/PhysRevLett.119.181301, arXiv:1705.06655 [astro-ph.CO]
10. Ballesteros, G., Redondo, J., Ringwald, A., Tamarit, C.: Standard model-axion-seesaw-Higgs portal inflation. Five problems of particle physics and cosmology solved in one stroke. JCAP

1708(08), 001 (2017). https://doi.org/10.1088/1475-7516/2017/08/001, arXiv:1610.01639 [hep-ph]

11. Ballesteros, G., Taoso, M.: Primordial black hole dark matter from single field inflation. Phys. Rev. D **97**(2), 023501 (2018). https://doi.org/10.1103/PhysRevD.97.023501, arXiv:1709.05565 [hep-ph]

12. Banks, T., Seiberg, N.: Symmetries and strings in field theory and gravity. Phys. Rev. D **83**, 084019 (2011). https://doi.org/10.1103/PhysRevD.83.084019, arXiv:1011.5120 [hep-th]

13. Bezrukov, F.L., Shaposhnikov, M.: The standard model Higgs boson as the inflaton. Phys. Lett. B **659**, 703 (2008). https://doi.org/10.1016/j.physletb.2007.11.072, arXiv:0710.3755 [hep-th]

14. Bilenky, S.M., Petcov, S.T., Pontecorvo, B.: Lepton mixing, mu –> e + gamma decay and neutrino oscillations. Phys. Lett. **67B**, 309 (1977). https://doi.org/10.1016/0370-2693(77)90379-3

15. Bjorken, J.D.: The November Revolution: a theorist reminisces. In: Presented at the SLAC Symposium on the Tenth Anniversary of the November Revolution, Stanford, California, 14 November 1984 (Fermilab-Conf-85/58)

16. Caldwell, A., Merle, A., Schulz, O., Totzauer, M.: Global Bayesian analysis of neutrino mass data. Phys. Rev. D **96**(7), 073001 (2017). https://doi.org/10.1103/PhysRevD.96.073001, arXiv:1705.01945 [hep-ph]

17. Carr, B.: Primordial black holes as dark matter and generators of cosmic structure. arXiv:1901.07803 [astro-ph.CO]

18. Chen, S.L., Frigerio, M., Ma, E.: Large neutrino mixing and normal mass hierarchy: a discrete understanding. Phys. Rev. D **70**, 073008 (2004). Erratum: [Phys. Rev. D **70**, 079905 (2004)]. https://doi.org/10.1103/PhysRevD.70.079905, https://doi.org/10.1103/PhysRevD.70.073008 [hep-ph/0404084]

19. Cheng, T.P., Li, L.F.: Gauge Theory of Elementary Particle Physics. Clarendon, Oxford, UK (1984)

20. Close, F.: A November revolution: the birth of a new particle. CERN Cour. **44**(10), 25 (2004)

21. Davis, R.: A review of the Homestake solar neutrino experiment. Prog. Part. Nucl. Phys. **32**, 13 (1994). https://doi.org/10.1016/0146-6410(94)90004-3

22. Davis, R.: Raymond Davis Jr. biographical. Autobiographical sketch for Nobel prize of 2002. https://www.nobelprize.org/prizes/physics/2002/davis/biographical/. Accessed 8 July 2019

23. Davoudiasl, H., Kitano, R., Li, T., Murayama, H.: The new minimal standard model. Phys. Lett. B **609**, 117 (2005). https://doi.org/10.1016/j.physletb.2005.01.026 [hep-ph/0405097]

24. Dell'Oro, S., Marcocci, S., Viel, M., Vissani, F.: Neutrinoless double beta decay: 2015 review. Adv. High Energy Phys. **2016**, 2162659 (2016). https://doi.org/10.1155/2016/2162659, arXiv:1601.07512 [hep-ph]

25. Drees, M., Xu, Y.: Critical Higgs inflation and second order gravitational wave signatures. arXiv:1905.13581 [hep-ph]

26. Ezquiaga, J.M., Garcia-Bellido, J., Ruiz Morales, E.: Primordial black hole production in critical Higgs inflation. Phys. Lett. B **776**, 345 (2018). https://doi.org/10.1016/j.physletb.2017.11.039, arXiv:1705.04861 [astro-ph.CO]

27. Fogli, G.L., Lisi, E., Mirizzi, A., Montanino, D.: Probing supernova shock waves and neutrino flavor transitions in next-generation water-Cerenkov detectors. JCAP **0504**, 002 (2005). https://doi.org/10.1088/1475-7516/2005/04/002 [hep-ph/0412046]

28. Fukuda, Y., et al. [Super-Kamiokande Collaboration]: Evidence for oscillation of atmospheric neutrinos. Phys. Rev. Lett. **81**, 1562 (1998). https://doi.org/10.1103/PhysRevLett.81.1562 [hep-ex/9807003]

29. Gariazzo, S., Archidiacono, M., de Salas, P.F., Mena, O., Ternes, C.A., Trtola, M.: Neutrino masses and their ordering: global data, priors and models. JCAP **1803**(03), 011 (2018). https://doi.org/10.1088/1475-7516/2018/03/011, arXiv:1801.04946 [hep-ph]

30. Gell-Mann, M., Ramond, P., Slansky, R.: Complex spinors and unified theories. Conf. Proc. C **790927**, 315 (1979). [arXiv:1306.4669 [hep-th]]

31. Glashow, S.L.: Partial symmetries of weak interactions. Nucl. Phys. **22**, 579 (1961). https://doi.org/10.1016/0029-5582(61)90469-2

32. Harlow, D., Ooguri, H.: Symmetries in quantum field theory and quantum gravity. arXiv:1810.05338 [hep-th]
33. Heavens, A.F., Sellentin, E.: Objective Bayesian analysis of neutrino masses and hierarchy. JCAP **1804**(04), 047 (2018). https://doi.org/10.1088/1475-7516/2018/04/047, arXiv:1802.09450 [astro-ph.CO]
34. Hill, H.: Lower limit on the half-life of neutrinoless double-beta decay. Phys. Today **15**, (2019). https://doi.org/10.1063/PT.6.1.20191015a
35. Hook, A.: TASI lectures on the strong CP problem and axions. arXiv:1812.02669 [hep-ph]
36. Jungman, G., Kamionkowski, M., Griest, K.: Supersymmetric dark matter. Phys. Rept. **267**, 195 (1996). https://doi.org/10.1016/0370-1573(95)00058-5 [hep-ph/9506380]
37. Kobayashi, M., Maskawa, T.: CP violation in the renormalizable theory of weak interaction. Prog. Theor. Phys. **49**, 652 (1973). https://doi.org/10.1143/PTP.49.652
38. Laudan, L., Leplin, J.: Empirical equivalence and underdetermination. J. Philos. **88**, 449 (1991)
39. Lee, B.W., Pakvasa, S., Shrock, R.E., Sugawara, H.: Muon and electron number nonconservation in a V-A gauge model. Phys. Rev. Lett. **38**, 937 (1977). Erratum: [Phys. Rev. Lett. **38**, 1230 (1977)]. https://doi.org/10.1103/PhysRevLett.38.937, https://doi.org/10.1103/PhysRevLett.38.1230
40. Lei, M., Steinberg, N., Wells, J.D.: Probing non-standard neutrino interactions with supernova neutrinos at hyper-K. arXiv:1907.01059 [hep-ph]
41. Long, A.J., Raveri, M., Hu, W., Dodelson, S.: Neutrino mass priors for cosmology from random matrices. Phys. Rev. D **97**(4), 043510 (2018). https://doi.org/10.1103/PhysRevD.97.043510, arXiv:1711.08434 [astro-ph.CO]
42. Marciano, W.J., Sanda, A.I.: Exotic decays of the muon and heavy leptons in gauge theories. Phys. Lett. **67B**, 303 (1977). https://doi.org/10.1016/0370-2693(77)90377-X
43. Minkowski, P.: $\mu \rightarrow e\gamma$ at a rate of one out of 10^9 muon decays? Phys. Lett. **67B**, 421 (1977). https://doi.org/10.1016/0370-2693(77)90435-X
44. Mohapatra, R.N., Senjanovic, G.: Neutrino mass and spontaneous parity nonconservation. Phys. Rev. Lett. **44**, 912 (1980). https://doi.org/10.1103/PhysRevLett.44.912
45. Motohashi, H., Hu, W.: Primordial black holes and slow-roll violation. Phys. Rev. D **96**(6), 063503 (2017). https://doi.org/10.1103/PhysRevD.96.063503, arXiv:1706.06784 [astro-ph.CO]
46. Orlofsky, N., Pierce, A., Wells, J.D.: Inflationary theory and pulsar timing investigations of primordial black holes and gravitational waves. Phys. Rev. D **95**(6), 063518 (2017). https://doi.org/10.1103/PhysRevD.95.063518, arXiv:1612.05279 [astro-ph.CO]
47. Particle Data Group (Hikasa, K., et al.): Review of particle properties. Phys. Rev. **D45**, S1 (1992)
48. Particle Data Group (Montanet, L., et al.): Review of particle properties. Phys. Rev. **D50**, 1173 (1994)
49. Peccei, R.D., Quinn, H.R.: CP conservation in the presence of instantons. Phys. Rev. Lett. **38**, 1440 (1977). https://doi.org/10.1103/PhysRevLett.38.1440
50. Pinch, T.: Confronting Nature: The Sociology of Solar-Neutrino Detection. D. Reidel, Boston (1986)
51. Preskill, J., Wise, M.B., Wilczek, F.: Cosmology of the invisible axion. Phys. Lett. B **120**, 127 (1983) [Phys. Lett. **120B**, 127 (1983)]. https://doi.org/10.1016/0370-2693(83)90637-8
52. Rubio, J.: Higgs inflation. Front. Astron. Space Sci. **5**, 50 (2019). https://doi.org/10.3389/fspas.2018.00050, arXiv:1807.02376 [hep-ph]
53. Salam, A.: Weak and electromagnetic interactions. Conf. Proc. C **680519**, 367 (1968)
54. Schechter, J., Valle, J.W.F.: Neutrino masses in $SU(2) \times U(1)$ theories. Phys. Rev. D **22**, 2227 (1980). https://doi.org/10.1103/PhysRevD.22.2227
55. Schirber, M.: WIMP alternatives come out of the shadows. APS Phys. https://physics.aps.org/articles/v11/48. Accessed 8 July 2019
56. Schwetz, T., Freese, K., Gerbino, M., Giusarma, E., Hannestad, S., Lattanzi, M., Mena, O., Vagnozzi, S.: Comment on "Strong evidence for the normal neutrino hierarchy". arXiv:1703.04585 [astro-ph.CO]

57. Shaposhnikov, M.: The Higgs boson and cosmology. Philos. Trans. R. Soc. Lond. A **373**(2032), 20140038 (2015). https://doi.org/10.1098/rsta.2014.0038
58. Simpson, F., Jimenez, R., Pena-Garay, C., Verde, L.: Strong Bayesian evidence for the normal neutrino hierarchy. JCAP **1706**(06), 029 (2017). https://doi.org/10.1088/1475-7516/2017/06/029, arXiv:1703.03425 [astro-ph.CO]
59. Tanner, N.P.: Decrees of the Ecumenical Councils, 2 vols. Georgetown University Press (1990)
60. Trimble, V., Reines, F.: The solar neutrino problem—a progress(?) report. Rev. Mod. Phys. **45**, 1 (1973). https://doi.org/10.1103/RevModPhys.45.1
61. Valle, J.W.F.: Neutrino physics from A to Z: two lectures at Corfu. PoS CORFU **2016**, 007 (2017). https://doi.org/10.22323/1.292.0007, arXiv:1705.00872 [hep-ph]
62. Weinberg, S.: A model of leptons. Phys. Rev. Lett. **19**, 1264 (1967). https://doi.org/10.1103/PhysRevLett.19.1264
63. Wells, J.D.: Beyond the hypothesis: theory's role in the genesis, opposition, and pursuit of the Higgs boson. Stud. Hist. Philos. Sci. B **62**, 36 (2018). https://doi.org/10.1016/j.shpsb.2017.05.004
64. Wells, J.D.: Discovery goals and opportunities in high energy physics: a defense of BSM-oriented exploration over signalism. arXiv:1904.02769 [physics.hist-ph]
65. Winter, W.: Lectures on neutrino phenomenology. Nucl. Phys. Proc. Suppl. **203–204**, 45 (2010). https://doi.org/10.1016/j.nuclphysbps.2010.08.005, arXiv:1004.4160 [hep-ph]
66. Yanagida, T.: Horizontal gauge symmetry and masses of neutrinos. Conf. Proc. C **7902131**, 95 (1979)

Printed in the United States
By Bookmasters